ASCE Manuals and Reports on Engineering Practice No. 138

Structural Fire Engineering

Prepared by the Fire Protection Committee of the Structures Engineering Institute of the American Society of Civil Engineers and Edited by Kevin J. LaMalva, P.E.

ASCE AMERICAN SOCIETY OF CIVIL ENGINEERS

SEI STRUCTURAL ENGINEERING INSTITUTE

Published by the American Society of Civil Engineers

Library of Congress Cataloging-in-Publication Data

Names: LaMalva, Kevin J., editor. | Structures Engineering Institute
(American Society of Civil Engineers). Fire protection Committee, editor.
Title: Structural fire engineering / prepared by the Fire Protection Committee of the
Structures Engineering Institute of the American Society of Civil Engineers; edited by
Kevin J. LaMalva, P.E.
Description: Reston, Virginia: American Society of Civil Engineers, [2018] |
Series: ASCE manuals and reports on engineering practice; no. 138 |
Includes bibliographical references and index.
Identifiers: LCCN 2018024125 | ISBN 9780784415047 (hardcover: alk. paper) |
ISBN 9780784481493 (pdf) | ISBN 9780784481714 (epub)
Subjects: LCSH: Building, Fireproof.
Classification: LCC TH1065 .S77 2018 | DDC 693.8/2—dc23
LC record available at https://lccn.loc.gov/2018024125

Published by American Society of Civil Engineers
1801 Alexander Bell Drive
Reston, Virginia 20191-4382
www.asce.org/bookstore | ascelibrary.org

Errata: **Errata, if any, can be found at https://doi.org/10.1061/9780784415047.**

ISBN 978-0-7844-1504-7 (print)
ISBN 978-0-7844-8149-3 (PDF)
ISBN 978-0-7844-8171-4 (ePub)
Manufactured in the United States of America.

24 23 22 21 20 19 18 1 2 3 4 5

MANUALS AND REPORTS ON ENGINEERING PRACTICE

(As developed by the ASCE Technical Procedures Committee, July 1930, and revised March 1935, February 1962, and April 1982)

A manual or report in this series consists of an orderly presentation of facts on a particular subject, supplemented by an analysis of limitations and applications of these facts. It contains information useful to the average engineer in his or her everyday work, rather than findings that may be useful only occasionally or rarely. It is not in any sense a "standard," however; nor is it so elementary or so conclusive as to provide a "rule of thumb" for nonengineers.

Furthermore, material in this series, in distinction from a paper (which expresses only one person's observations or opinions), is the work of a committee or group selected to assemble and express information on a specific topic. As often as practicable the committee is under the direction of one or more of the Technical Divisions and Councils, and the product evolved has been subjected to review by the Executive Committee of the Division or Council. As a step in the process of this review, proposed manuscripts are often brought before the members of the Technical Divisions and Councils for comment, which may serve as the basis for improvement. When published, each work shows the names of the committees by which it was compiled and indicates clearly the several processes through which it has passed in review, so that its merit may be definitely understood.

In February 1962 (and revised in April 1982), the Board of Direction voted to establish a series titled "Manuals and Reports on Engineering Practice," to include the Manuals published and authorized to date, future Manuals of Professional Practice, and Reports on Engineering Practice. All such Manual or Report material of the Society would have been refereed in a manner approved by the Board Committee on Publications and would be bound, with applicable discussion, in books similar to past Manuals. Numbering would be consecutive and would be a continuation of present Manual numbers. In some cases of joint committee reports, bypassing of Journal publications may be authorized.

A list of available Manuals of Practice can be found at http://www.asce.org/bookstore.

CONTENTS

PREFACE

Uncontrolled fire within a contemporary building is an extraordinary event that can have severe consequences. When structural systems are heated by fire, they experience thermal effects that are not contemplated by conventional structural engineering design. Since the early 20th century, building codes have primarily adopted standard fire resistance design. This design option is based on furnace testing of isolated structural components and relies almost exclusively on insulation for structural fire protection. Using this approach, designers select qualified assemblies from available listings to meet prescribed levels of fire resistance. As an alternative approach, structural fire engineering design explicitly evaluates the performance of structural systems under fire exposure, in a similar manner as other design loads are treated in structural engineering practice.

The 2016 edition of ASCE/SEI 7 (*Minimum Design Loads and Associated Criteria for Buildings and Other Structures*) includes a new Chapter 1, Section 1.3.7, "Fire Resistance." In addition to being the first time that fire resistance has ever been addressed in this standard, this section is intended to commence a new industry-consensus standard of care for structural fire protection. The default option is for the designer to strictly adhere to the requirements and restrictions of standard fire resistance design per the applicable building code without exception or extrapolation. As an alternative, ASCE/SEI 7-16 permits the designer to adopt a structural fire engineering approach as constituted in the standard's new Appendix E, "Performance-Based Design Procedures for Fire Effects on Structures." Notably, the inclusion of Appendix E in ASCE/SEI 7 marks the first time that fire effects are considered as an explicit design load in a U.S. structural engineering standard.

The main objective of the manual is to provide best practices when a structural fire engineering design approach is adopted, as permitted by ASCE/SEI 7 Chapter 1, Section 1.3.7, and as constituted in Appendix E. Because standard fire resistance design does not contemplate structural system performance or explicit performance objectives, there is no practical method for a designer to quantitatively compare the level of safety provided by a structural fire engineering design to that provided by a standard fire resistance design. Furthermore, it is not reasonable to require that a structural fire engineering design be "equivalent" to a standard fire resistance design, which does not provide any affirmative quantification of structural system performance under fire exposure. Hence, the best practices described in this manual for successful implementation of a structural fire engineering design are intended to stand independently from acceptance criteria in standard fire resistance design. Although the manual was written with the aim of providing best practices for engineers, it is hoped that it will also be useful to architects, building officials, and academics concerned with performance-based design for structural fire safety. Also, the combination of ASCE/SEI 7 Appendix E, this manual, and other related standards should advance structural fire engineering design as a distinct discipline within structural engineering.

Kevin J. LaMalva, P.E.
Structural Fire Engineer
Simpson Gumpertz & Heger Inc.

ACKNOWLEDGMENTS

Contributions to the manual were received from consultants, professional researchers, and academic researchers. This manual has been prepared by the ASCE/SEI Fire Protection Committee, which included the following members at the time this manual was developed:

K. LaMalva, *Chair*
M. Garlock, *Past Chair*
F. Alfawakhiri
D. Banerjee
E. Beneberu
L. Bisby
N. Braxtan
M. Brightman
A. Carlton
D. Choe
S. Deeny
E. Fischer
J. Gales
J. Glassman
Q. Guo
E. Hantouche
R. Imani
A. Jeffers
E. Jensen
C. Jones
A. Jowsey
M. Ketabdar
M. Mahamid
H. Mahmoud
C. Maluk
T. McAllister
M. Morovat
K. Mueller
S. Quiel
H. Quiquero
M. Seif
S. Selamet
C. Smith
M. Smith
M. Soliman
R. Sun
S. Szoke
C. Zhang
A. Zhou

The following Committee Task Groups made major contributions to the manual:

Structural Analysis Task Group

Therese McAllister, *Co-Leader*
Spencer Quiel, *Co-Leader*
Matthew Brightman
Erica Fischer
Elie Hantouche
Reza Imani
Mustafa Mahamid
Hussam Mahmoud
Ruirui Sun

Mechanical Properties Task Group

Ann Jeffers, *Co-Leader*
Ali Morovat, *Co-Leader*
Farid Alfawakhiri
Dilip Banerjee
Eyosias Benebera
Qianru Guo
Alin Jensen
Mina Seif
Steven Szoke

Acceptance Criteria Task Group

John Gales, *Co-Leader*
Aixi Zhou, *Co-Leader*
Luke Bisby
Nicole Braxtan
Susan Deeny
Jonathan Glassman
Cristian Maluk
Hailey Quiquero
Serdar Selamet
Christopher Smith
Matthew Smith

The committee would like to thank Maria Garlock, who served as chair during the development of ASCE/SEI 7 Appendix E. Dr. Garlock's previous leadership of this committee deserves special acknowledgment.

The manual was reviewed by an ASCE/SEI Blue Ribbon Panel, consisting of the following members:

Lisa Choe
Richard Davis
Michael Engelhardt
Ramon Gilsanz
Nestor Iwankiw
Chris Jelenewicz
Venkatesh Kodur
Panagiotis Kotsovinos
Brian Kukay
Brian Meacham
Bernie Till
Jose Torero
Amit Varma

The contributions of the following individuals who are not committee members are appreciated:

Brad Douglas
Morgan Hurley
William Luecke
Hossein Mostafaei
Mohammed Sultan

The efforts of the following individuals who assisted the committee obtain permissions are appreciated:

Jillian Aberdale
Joan Cunningham
Michelle Howard

The cover image was captured at the National Fire Research Laboratory (NFRL) of the National Institute of Standards and Technology (NIST), where real-scale structural-fire testing is conducted to help advance structural fire engineering in the United States.

PART 1

INTRODUCTION TO STRUCTURAL FIRE ENGINEERING DESIGN

CHAPTER 1
INTRODUCTION AND KEY TERMINOLOGY

1.1 OVERVIEW

Heating and subsequent cooling of structural systems during fire exposure causes thermal load effects and nonlinear responses in structures, as well as temperature-dependent changes in material properties. Fire effects from restrained thermal expansion or contraction (if occurring) may result in significant loads and bending moments on members and connections. Reduced material stiffness and strength, and any other irreversible effects from fire exposure, may also result in large member deflections and deformations.

Fire protection of buildings may be achieved through the use of active and/or passive systems. Automatic fire sprinkler systems are commonly used for active fire protection. Fire sprinklers have demonstrated to be effective to control or extinguish fires during their early stages of growth approximately 85% to 90% of the time, where the fire was large enough to activate the sprinkler system and the sprinklers were present in the area of the fire (Ahrens 2017). Otherwise, fire sprinkler systems have failed to control the fire. The frequency of major building fires is relatively low because of (1) the small probability of ignition and the small probability of a fire reaching flashover, (2) occupant or fire department intervention, and/or (3) automatic sprinkler system control of the fire before it becomes fully developed. However, certain events (e.g., terrorism, arson, fires in multiple areas simultaneously, poor system maintenance, inadequate system design, building under construction, and so forth) may result in uncontrolled fires that are structurally significant. In these rare cases, the structural system may reach elevated temperatures for a significant time.

Passive fire protection systems provide the first line of defense in buildings without sprinklers, and a second line of defense in sprinklered buildings during significant fire events by delaying heating of structural elements, but they cannot prevent heating or render any object *fireproof*. Passive systems include the structural systems themselves and any materials applied directly to structural members, such as spray-applied, fire resistant materials (SFRMs), intumescent paints, and gypsum board enclosures. Fire and smoke barriers compartmentalize areas of a building to limit the spread of fire and smoke from the location of fire origin to other building spaces. Load-bearing elements that also function as fire resistant separation assemblies (e.g., floor slabs and bearing walls) may act as a thermal and flame barrier in addition to providing structural integrity.

This manual presents best practices for the performance-based design of structural systems for fire effects, referred to as *structural fire engineering design* (SFED) herein. Although this manual does not pertain to and should not be used for prescriptive design, referred to as *standard fire resistance design* (SFRD) herein, SFRD is discussed relative to SFED to provide context. Also, this manual is limited to analyses of structural systems that are not significantly damaged by other hazard events, such as an earthquake or an explosive blast.

1.2 BACKGROUND

The fire protection of most structural systems is provided according to prescriptive methods in building codes, which specify fire resistance ratings for members or subassemblies, such as floor beams and columns or floor systems. Building codes specify levels of passive fire protection by types of construction, occupancy, and layout (area and height). Designers have the freedom to select construction materials and assemblies to meet the fire resistance requirements using approved methods, primarily by qualification testing, as discussed in, for example, ASTM E119 (ASTM 2016). Also, certain calculation methods that demonstrate equivalence to ASTM E119 fire-rated assemblies [e.g., Standard 29-05 (ASCE/SEI/SFPE 2005)] are permitted. These calculation methods primarily interpolate between available furnace test results.

Standard fire testing such as described in ASTM E119 provides comparative fire ratings for components and assemblies. Each test uses the same standard fire (time–temperature) curve, which continually rises in temperature. The structural components and assemblies are tested against an established set of failure criteria, which address heat transfer to structural members, integrity of passive fire protection systems, and deformation and stability of the mock-up assembly. The test is halted when any of the failure criteria is reached.

Furnaces for standard fire tests have finite dimensions that limit test specimen sizes. For instance, composite floor assemblies are typically tested for lengths up to nominally 17 ft (UL 2015). This specimen size can provide an accurate assessment of heat transfer through the floor, but it does not adequately simulate the structural response of actual structural systems. Furthermore, standard fire testing does not provide performance evaluation of components or assemblies for realistic fire conditions.

Building codes generally allow for the use of alternative means and methods with the approval of the building official so that innovative methods or materials are not excluded. SFED is an alternative method that allows for explicit evaluation of structural performance under realistic fire conditions. The International Code Council (ICC) developed the *2003 ICC Performance Code for Buildings and Facilities* to provide guidance for performance-based design (ICC 2012b). Also, the SFPE *Engineering Guide to Performance-Based Fire Protection* (SFPE 2007) outlines a process for using a performance-based approach for building fire safety. Last, ASCE 7-16 (ASCE/SEI 2016) addresses performance-based design for buildings and other structures through provisions for alternate means and methods.

Much of the impetus for performance-based evaluation of structural system response to fire effects is the result of the fire-induced structural damage and collapses at the World Trade Center (WTC) complex in New York City on September 11, 2001 (Figures 1-1 and 1-2). However, the performance of other buildings, such as WTC 5, during significant fire events also supports the need for structural engineering participation in structural fire protection

Figure 1-1. WTC 1 and 2 sites after collapse from fire.
Source: NOAA (2001).

Figure 1-2. WTC 7 fire event.
Source: Wikipedia, https://en.wikipedia.org/wiki/7_World_Trade_Center.

design. The following steel-framed buildings, each with hourly ratings of fire resistance for floor beams and columns, and subject to uncontrolled burning of ordinary office building contents, had varying responses:

- The First Interstate Bank building in Los Angeles was built in 1973 with lightweight composite slab and moment frame construction. An uncontrolled fire burned from the 12th to the 16th floors in 1988. The primary structural frame was undamaged, and the building was repaired (Routley 1988, Nelson 1989).
- One Meridian Plaza of Philadelphia was built in 1969 with noncomposite floors and moment frame construction. Severe fires between the 21st and 29th floors in 1991 resulted in structural damage to floor sections, with some sagging as much as 3 ft (Figure 1-3), and cracking through the reinforced concrete floors in many places, but without significant structural failures or obvious damage to the columns (Routley et al.1991). However, because of the extent of secondary damage, the building was demolished.
- WTC 5 in New York City was built in 1970 with lightweight composite floors and moment frame construction, which included field-bolted connections between the floor beams and interior column trees. On September 11, 2001, uncontrolled fires over most of its nine stories led

Figure 1-3. One Meridian Plaza building after fire.
Source: Routley et al. (1991).

to interior failure of large sections of floor over four floors in areas not damaged by falling debris from WTC 1 (Figure 1-4). The failures were attributed to bolt tear-out at beam-to-column tree connections (FEMA 2002, LaMalva et al. 2009). The building was demolished.

- WTC 7 in New York City was built in 1985 with lightweight composite floors and exterior moment frame construction. The floor framing on the east side of WTC 7, where the collapse initiated, had long-span floor beams (52 ft) that framed into the girders on one side only; no significant restraint of thermal expansion by the floor beams was provided by the interior girders. On September 11, 2001, uncontrolled fires over six floors between the 7th and 14th stories led to failure of the girder's seated connections because of thermal expansion of the floor beams, which contributed to the global collapse of the entire building (McAllister et al. 2008).

Concrete buildings are also subject to damage or collapse during significant fire events; notable failures include the following:

- The Military Personnel Record Center was a large six-story office building located in Overland, Missouri, that was constructed of reinforced concrete without expansion joints in the late 1950s. In 1973, a fire on the sixth floor burned for 20 hours from the high fuel load of the record files. After approximately 12 hours, 30% of the roof collapsed (Figure 1-5), followed by failure of most interior columns on the sixth floor (Sharry et al. 1974). The collapse and damage were attributed to thermal expansion of the 7 in.-thick reinforced concrete roof slab (Beitel and Iwankiw 2002); lateral roof displacements of up to 23 in. occurred. After the sixth floor was removed, the building continues to be in use.

Figure 1-4. WTC 5 after fire.
Source: FEMA (2002).

Figure 1-5. Military Personnel Record Center after fire.
Source: Sharry et al. (1974).

- Katrantzos Sport department store in Athens, Greece, was an eight-story, reinforced concrete building. A fire started at the seventh floor in 1980 and rapidly spread because of lack of vertical or horizontal compartmentalization. A major portion of the fifth through eighth floors collapsed, as well as various other floor and column failures throughout the building (Papaioannou 1986). The cause of these failures was attributed to restraint of thermal expansion that overloaded specific elements and connections (Beitel and Iwankiw 2002). The building was demolished.

- The Delft Technical University Faculty of Architecture building, Delft, Netherlands, was completed in 1970 as a combination of six 3-story steel structures and a 13-story reinforced concrete building. In 2008, an uncontrolled fire from the 6th to 11th floors resulted in the collapse of the north side of the 13-story building (Figure 1-6). Loss of cross section, presumably from spalling, was observed in column sections (Meacham et al. 2010). The cause of the collapse has not been determined, but fire-induced spalling appears to be a contributing factor. The building was demolished.

The disparity in the structural performance of these buildings illustrates the lack of predictability during significant fire events when the performance of structural systems under fire conditions is not explicitly evaluated. Whereas most buildings perform satisfactorily in significant fires, for example, the Echelon I building (Blackmore 2010), SFRD methods cannot predict the performance of a structural system in fire conditions. This is particularly critical for steel-framed buildings with long spans subject to forces associated with restraint of thermal expansion and contraction (Bailey et al. 1999), as well as timber buildings because of the structure's contribution to the overall fuel load. Structural damage and collapses of contemporary buildings due to fire, such as One Meridian Plaza building (Eisner and Manning 1991), WTC 1 and 2 buildings (Gross and McAllister 2005), WTC 5 building (LaMalva et al. 2009), WTC 7 building (McAllister et al. 2008), Windsor Tower (Fletcher et al. 2006), Plasco building (Taylor 2017), and Delft

Figure 1-6. Delft Technical University Faculty of Architecture building. Source: Broekbakema: Architecture, Interior and Masterplanning, reproduced with permission.

Technical University building (Meacham et al. 2010) highlight the need for structural engineering participation in evaluating fire effects on structural systems.

KEY TERMINOLOGY

The following key terminology applies to the information presented in this manual:

Active Fire Protection System: A fire protection system designed to sense, control, or suppress fire, which requires activation by sensing fire by-products (e.g., hot gases). An automatic fire sprinkler system is an example of an active fire protection system.

Alternative Design: An option within a code whereby compliance is achieved by demonstrating that a proposed design will satisfy required performance objectives using engineering analysis.

Automatic Fire Sprinkler System: An active fire suppression system or control device that operates automatically when its heat-activated element reaches its thermal rating or above, allowing water to discharge over a specified area.

Available Safe Egress Time (ASET): The available time for building occupants to evacuate or reach a place of safety prior to the onset of untenable conditions or inadequate structural performance. Evaluation of tenable conditions is not within the scope of this manual and is addressed separately from SFED.

Building Refuge Area: Any location within a building that is designed to safely shelter occupants during a fire event, when immediate evacuation may not be safe or possible.

Burnout: Extinguishment of fire due to the consumption of fuel load.

Coefficient of Thermal Expansion: Measure of the amount that an element's size changes with temperature. Common units for the coefficient of thermal expansion are mm/mm/K.

Conductive Heat Transfer: Transfer of heat through a solid body, such as a concrete wall, due to a spatial thermal gradient often measured in kilowatts per square meter (kW/m^2).

Convective Heat Transfer: Transfer of heat due to bulk motion of a fluid to a solid surface, such as hot gases moving along a concrete wall; often measured in kilowatts per square meter (kW/m^2).

Density: Mass of a material per unit volume. Common units for density are kilograms per cubic meter (kg/m^3).

Emissivity: Ratio of the radiation emitted by a surface to the radiation emitted by a blackbody (ideal emitter) at the same temperature.

Enclosed Compartment: A room with intersecting ceiling, floor, and wall surfaces that may have openings.

Fire: An oxidation process that results in the burning of flammable materials and produces heat.

Fire Barrier: A fire resistance–rated assembly designed to restrict the spread of fire when assembly integrity is maintained.

Fire Effects: Thermal and structural response due to fire exposure and subsequent cooling.

Fire Exposure: The extent to which materials, products, or assemblies are subjected to the conditions created by fire.

Fire Model: A physical or mathematical representation of the dynamics of burning and associated processes.

Fire Resistance: The time duration that a mock-up assembly is able to withstand furnace heating based upon standard fire testing requirements and acceptance criteria.

Fire Resistance–Rated Assembly: An assembly that provides a specified level of fire resistance (e.g., two hours) used as part of a passive fire protection system. Structural members thermally insulated with SFRMs or fire barrier walls are examples of fire resistance–rated assemblies.

Fire Scenario: A description or set of conditions that defines the development of fire, such as ventilation area and compartment geometry, and the distribution of fuel load in the compartment or building area of interest.

Flashover: A rapid transition in a compartment fire from localized burning to burning of all combustible materials in a compartment. Flashover can only occur in an enclosed compartment with sufficient fuel and ventilation, where the ceiling can trap hot gases that lead to radiant heating of all fuels to the point of combustion.

Fuel Load: The total quantity of combustible contents within a building, space, or area.

Fully Developed Fire: The condition of a fire after flashover. This condition depends on the ventilation available to the fire or the mass loss rate of the fuel.

Heat Flux: The rate of heat transfer per unit area; commonly measured in kilowatts per square meter (kW/m^2).

Heat Release Rate: The rate at which the combustion reactions of a fire produce heat; commonly measured in megawatts (MW).

Heat Transfer: The exchange of thermal energy due to a temperature difference.

Intumescent Coatings: Paint or coating applied in layers as passive fire protection. When exposed to heat, intumescent coatings char and swell, increasing the coating volume and thickness and decreasing its density and thermal conductivity to create an insulating effect.

Localized Fire: A fire that burns combustibles at a given location and would heat a structure locally. Localized burning can occur in open areas, large spaces, areas with high ceilings, or other similar locations.

Openings: Doors, windows, or penetrations into a compartment that provide a path for ventilation to a fire.

Passive Fire Protection System: A fire protection system designed to contain or slow the spread of fire by means of building construction.

Radiative Heat Transfer: Transfer of heat through electromagnetic waves; often measured in kilowatts per square meter (kW/m^2).

Required Safe Egress Time (RSET): The required time from fire ignition to the time when building occupants have evacuated or reached a place of safety.

Residual Material Properties: The ambient thermal and/or mechanical properties of a construction material that has been exposed to elevated temperatures from fire and then returned to ambient temperature.

Specific Heat: Measure of the amount of energy required to raise the temperature of a unit of mass a unit of temperature; often measured in joules per kilogram-kelvin [J/(kg-K)].

Spray-Applied Fire Resistive Material (SFRM): A passive fire protection material intended for direct application to structural building members. The intent of this material is to increase the fire resistance of building members primarily through its insulating characteristics.

Standard Fire Curve: A temperature history used for standard fire resistance testing (e.g., ASTM E119 fire curve).

Standard Fire Resistance Design (SFRD): The selection of qualified fire resistive assemblies to meet code requirements for structural fire resistance (also known as prescriptive-based design). The rating of a fire resistive assembly is based on its performance in standard furnace testing per specific acceptance criteria.

Structural Design Fire: An uncontrolled fire that has the potential to affect the integrity and stability of a structure that is used for the design and evaluation of the structure.

Structural Fire Engineering Design (SFED): The explicit design of structural systems to adequately endure thermal load effects from structural design fires based on specific performance objectives.

Thermal Boundary Condition: The temperature and/or heat flux to which an assembly or the structure is subjected during or after fire exposure based upon the radiative and convective heating/cooling conditions at exposed surfaces.

Thermal Conductivity: Measure of a material's ability to conduct heat. Common units for thermal conductivity are watts per meter-kelvin W/(m-K).

REFERENCES

Ahrens, M. 2017. "U.S. experience with sprinklers." Quincy, MA: National Fire Protection Association.

ASCE/SEI (Structural Engineering Institute). 2016. *Minimum design loads and associated criteria for buildings and other structures*. Standard 7-16. Reston, VA: ASCE.

ASCE/SEI/SFPE (ASCE/Structural Engineering Institute/Society of Fire Protection Engineers). 2005. *Calculation methods for structural fire protection*. Standard 29-05. Reston, VA: ASCE.

ASTM International. 2016. *Standard test methods for fire tests of building construction and materials*. E119-16a. West Conshohocken, PA: ASTM International.

Bailey, C., T. Lennon, and D. Moore. 1999. "The behaviour of full-scale steel-framed buildings subjected to compartment fires." *Struct. Eng.* 77(8), 15–21.

Beitel, J., and N. Iwankiw. 2002. "Analysis of needs and existing capabilities for full-scale fire resistance testing." *GCR 02-843-1*. Gaithersburg, MD: NIST.

Blackmore, G. 2010. "Austin plane crash: Incident report and analysis." *Fire Eng*. 163(6), 55.

Eisner, H., and W. Manning. 1991. "One Meridian Plaza fire." *Fire Eng*. 144(8), 50–70.

FEMA. 2002. "World Trade Center building performance study: Data collection, preliminary observations, and recommendations." Washington, DC: FEMA.

Fletcher, I., A. Borg, N. Hitchen, and S. Welch. 2006. "Performance of concrete in fire: Review of the state of the art, with a case study of the Windsor Tower fire." *Proc., 4th Int. Workshop for Structures in Fire,* Vol. 2, 779–790. Aveiro, Portugal: Campus Universitário de Santiago.

Gross, J., and T. McAllister. 2005. "Federal building and fire safety investigation of the World Trade Center disaster: Structural fire response and probable collapse sequence, structural response of World Trade Center of the World Trade Center towers." *NCSTAR 1-6*. Gaithersburg, MD: NIST.

ICC (International Code Council). 2012b. *ICC performance code for buildings and facilities*. Falls Church, VA: International Code Council.

LaMalva, K. J., J. Barnett, and D. Dusenberry. 2009. "Failure analysis of the World Trade Center 5 building." *J. Fire Protect. Eng.* 19(4), 261–274.

McAllister, T., R. Gann, J. Averill, J. Gross, W. Grosshandler, J. Lawson, K. McGrattan, et al. 2008. "Federal building and fire safety investigation of the World Trade Center disaster: Structural fire response and probable collapse sequence of World Trade Center Building 7." *NCSTAR 1–9*. Gaithersburg, MD: NIST.

Meacham, B., H. Park, M. Engelhardt, A. Kirk, V. Kodur, I. J. Straalen, et al. 2010. "Fire and collapse, Faculty of Architecture building, Delft University of Technology: Data collection and preliminary analyses." Accessed May 8, 2018. http://www.dr.dk/NR/rdonlyres/FECDDBA8-7428-460D-846F-0C3D424D3A81/2748348/Delft_brandrapport.pdf.

Nelson, H. 1989. "An engineering view of the fire at the first interstate bank building." *IR 89-4061*. Gaithersburg, MD: NIST.

NOAA (National Oceanic and Atmospheric Administration). 2001. "NOAA conducts more flights over World Trade Center site." Accessed May 8, 2018. http://www.noaanews.noaa.gov/stories/s798.htm.

Papaioannou, K. 1986. "The conflagration of two large department stores in the centre of Athens." *Fire Mater.* 10(3-4), 2011–2177.

Routley, G. 1988. "First Interstate Bank building fire, Los Angeles, California, May 4, 1988." *TR-022*. Emmitsburg, MD: U.S. Fire Administration.

Routley, G., C. Jennings, and M. Chubb. 1991. "High-rise office building fire: One Meridian Plaza, Philadelphia, Pennsylvania, February 23, 1991." *TR-049*. Emmitsburg, MD: U.S. Fire Administration.

SFPE (Society of Fire Protection Engineers). 2007. *Engineering guide to performance-based fire protection,* 2nd Ed. Bethesda, MD: Society of Fire Protection Engineers.

Sharry, J., C. Culver, R. Crist, and J. Hillelson. 1974. "Military personnel records center fire, Overland, MO, Part 1." *NFPA Fire J.* 68(3), 5–9.

Taylor, A. 2017. "Iranian high-rise collapses in huge fire, killing dozens." Accessed May 8, 2018. https://www.theatlantic.com/photo/2017/01/iranian-high-rise-collapses-in-huge-fire-killing-dozens/513671/.

UL (Underwriters Laboratories). 2015. *UL fire resistance directory,* Vol. 1. Northbrook, IL: Underwriters Laboratories.

CHAPTER 2
DESIGN APPROACHES

2.1 OVERVIEW

Structural performance under fire exposure can be expressed either in terms of fire resistance or fire effects. Two design philosophies are used when establishing fire resistance and fire effects: the first is based on measures of time and the second by assessing the evolution of strength. *Time* is conventionally used for SFRD requirements in U.S. building codes in which a particular assembly is shown to provide adequate resistance to a standard fire exposure under furnace test conditions for a period of time per specific acceptance criteria. *Strength* is used for SFED by comparing applied gravity loads and fire effects (e.g., thermally induced forces) to the evolving strength of the structural system.

In the context of structural engineering, the fundamental philosophy used to design structures can be expressed as a quantitative comparison: capacity > demand (i.e., the capacity shall exceed the demand). Also, design factors are applied to the capacity and demand as part of an overall reliability-based framework. *Demand* refers to loads that are imposed on a structure or that are generated by the structure, including its self-weight. *Capacity* refers to the global and local ability of a structure to carry an imposed *demand*. The design of a structure evaluates *demand* and *capacity* with respect to specific performance metrics including strength, stability, and stiffness.

Structural fire resistance based on measures of *time* primarily endeavors to control the *demand* on the structure due to heating, usually by means of protective insulation. Consequently, this design philosophy does not directly account for the *capacity* of the structure itself to endure fire exposure nor does it evaluate all aspects of the *demand* (e.g., thermally induced forces). With measures of *time*, thermally based failure criteria for standard fire tests

Table 2-1. Consideration of Demand and Capacity.

	Demand		Capacity
	Heating	Induced Forces	Structural Endurance
Standard Fire Resistance Design	Measured	Indeterminate	Indeterminate
Structural Fire Engineering	Calculated	Calculated	Calculated

are often the basis for determining the fire resistance. Such an approach does not evaluate *demand* with respect to specific performance objectives. Evaluation of structural fire performance in terms of *strength* accounts for all contributions of the *demand* on a structure, as well as the *capacity* of the structure to endure these demands with respect to required performance objectives. *Hence, there is no practical method for a designer to quantitatively compare the level of safety provided by an SFED to that provided by an SFRD* (Table 2-1).

2.2 STANDARD FIRE RESISTANCE DESIGN (SFRD)

SFRD is based on measures of *time* given a defined heating exposure. Specifically, fire resistance is defined as an hourly rating based on the results of standard fire testing. SFRD has specific requirements that describe the process or details of construction for the purpose of achieving an intended level of safety or performance. Standard time–temperature curves used for standard fire testing are included in ASTM E119-16a (2016) and ANSI/UL 263 (2011); international standards include ISO 834 (1999), EN 1363 (CEN 1999), and BS 476-20 (1987). There are also standard fire curves for rapid-rise or hydrocarbon pool fires, such as ASTM E1529 (2010) and UL 1709 (2011), which are used for special applications (e.g., tunnels).

Dating back as early as the 1927 Uniform Building Code (UBC), the level of fire resistance required for individual components of a structural system has been prescribed by building codes in the U.S. The prescribed level of fire resistance has generally been based on the building occupancy, height, and area (Goode 2004). Of note, the 1927 UBC required that columns and floors be 4-h and 3-h fire resistance rated, respectively, for buildings taller than 8 stories or 85 ft. (UBC 1927). More recently, U.S. building codes have allowed for specific reductions in fire resistance if the building is equipped with an automatic sprinkler system (ICC 2012). Within this framework, evaluation of structural systems under fire exposure reduces to the selection of qualified assemblies from available listings to meet prescribed levels of fire resistance.

2.2.1 Standard Fire Testing

Standard fire testing (e.g., ASTM E119) provides a prescriptive method of rank ordering through comparative testing of different structural assemblies under controlled furnace conditions. Typically, the structural assemblies are tested with applied fire protection (e.g., SFRM). Each test uses the same time–temperature curve, which continually rises in temperature, to heat structural components and assemblies against an established set of thermal and structural failure criteria. The standard fire exposure does not represent actual fire exposures (as illustrated in Figure 2-1). Standard fire testing was first introduced prior to 1920 and has remained relatively unchanged to the present day.

The standard fire test was developed on the assumption that the test assembly is representative of field construction (NFPA 1928, and Nelson 2003). However, because of size limitations of furnaces, members and assembly sizes are limited. For instance, floor assemblies are typically tested at spans no longer than 17 ft (5.2 m) (UL 2015), whereas an actual floor span may be much greater. Figure 2-2 shows a standard fire test of protected open

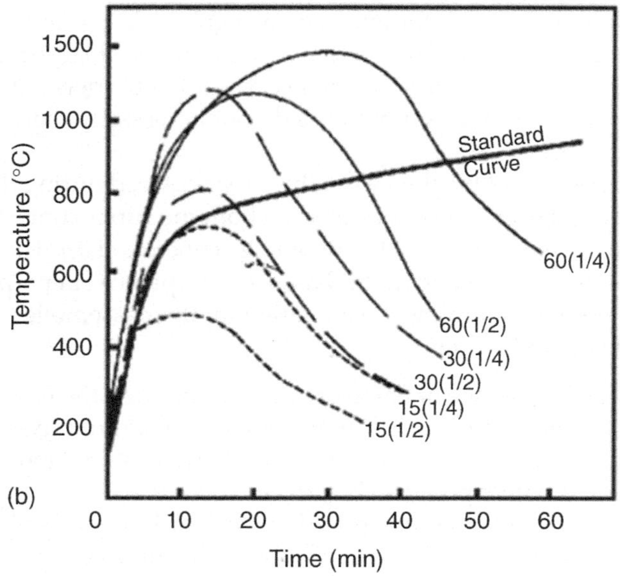

Figure 2-1. Furnace exposure versus actual fire exposures (fuel load in kg/m² and ventilation as a fraction of one wall).
Source: Lamont (2001), reproduced with permission.

Figure 2-2: Furnace test of protected open web steel joists.
Source: Kevin LaMalva, reproduced with permission.

web steel joists. The standard fire test evaluates individual components, as shown in Figure 2-3, but not structural systems.

The application of the standard fire test to floor assemblies is summarized here to illustrate key points regarding the (1) criteria used to determine a fire resistance rating, (2) the conditions of restraint to thermal expansion that may be applied, and (3) the application of gravity loads. Tests for beams, floor assemblies, columns, load-bearing walls, and nonload-bearing walls have their own set of criteria. For instance, some fire barriers have fire hose stream requirements, and column tests do not necessarily require applied loading.

The temperatures of the floor assembly are measured during the furnace exposure using thermocouples located on both the supporting steel members and the top and bottom of the floor (e.g., concrete slab). The assembly is given a rating, expressed in hours, based on acceptance or end-point criteria. Criteria used to determine the fire rating for floor assemblies per ASTM E119 include the following:

- Limiting the passage of flame or hot gases through the floor assembly such that the applied loads, which are the maximum load condition for the structural design criteria, are supported without igniting cotton waste on the unexposed slab surface,
- Temperatures on the unexposed surface of the floor slab shall not increase more than 250°F (121°C) above the initial temperature,
- Temperatures of *restrained* assemblies (with structural steel members) shall not have an average temperature greater than 1100°F (593°C), and

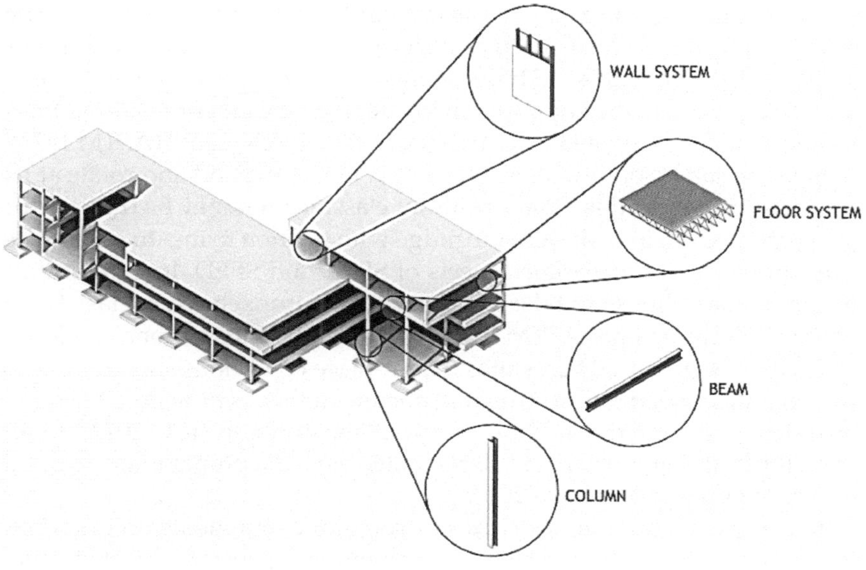

Figure 2-3. Structural system decomposition for furnace testing.
Source: LaMalva (2017a), reproduced with permission.

- Temperatures of conventional reinforced concrete structural members shall not have an average temperature of the tension steel at any section greater than 800°F (427°C) for cold-drawn prestressing steel and 1100°F (593°C) for reinforcing steel.

Failure criteria usually varies between different assembly types (e.g., floor assemblies versus columns). In all cases, the fire rating of a member or assembly is based on the first failure criterion reached. Fire resistance directories provide the hourly ratings for tested assemblies, but they do not provide information about the failure criterion upon which the listing is based. Nonetheless, exceedance of temperature limits usually governs.

ASTM first introduced the *restrained versus unrestrained* concept in the 1970s (ASTM 1971) based on the observation that thermal restraint provided by the furnace enclosure generally enhances the performance of assemblies during furnace testing. An assembly is considered *restrained* if it bears directly against the edges of the furnace at the outset of the test (ASTM 2016). In a furnace test, a beam and concrete slab assembly is considered *restrained* if both components horizontally bear directly against the edges of the furnace at the outset of the test; the assembly is considered *unrestrained* if both ends of the beam and slab are allowed to thermally expand without contacting the furnace edge. Many listings in the *UL Fire Resistance Directory* (UL 2015) permit reduced fire protection thicknesses to achieve certain fire resistance ratings if the designer can demonstrate that the assembly will be *restrained* when it is constructed as part of an actual structural system.

Appendix A4 of ASTM E119-73 defines a *restrained* condition as "one in which expansion at the support of a load-carrying element resulting from the effects of fire is resisted by forces external to the element" (ASTM 1973). In an *unrestrained* condition, the element is free to expand and rotate at its supports. However, this binary restraint classification may be difficult (or even paradoxical) for a designer to judge with relation to in-situ conditions considering the incompatible aspects of SFRD and SFED. In light of this, designers may choose to take a conservative approach when classifying restraint conditions per ASTM E 119. Notably, the International Building Code (IBC) Section C703.2.3 states that in-place conditions should be considered unrestrained unless structural documentation is provided that demonstrates a restrained condition in actual construction (ICC 2012). In all cases, the building official may be consulted as to the proper interpretation for a given project (LaMalva 2017b).

In actual construction, restraint of structural components occurs when the surrounding structural system resists thermal expansion of heated structural components, as well as subsequent thermal contraction during cooling. Restraint of thermal contraction is not accurately represented during standard fire testing. Many factors affect restraint conditions that may increase or decrease structural endurance to fire effects. For instance, a heated floor beam within a structural system may undergo thermal expansion and impose lateral loads on the girder and column connections. Depending on the characteristics of the connections, the thermal expansion may be resisted by inducing axial loads or moments until the thermally induced loads exceed either the beam or support capacity. Connections may also resist subsequent thermal contraction during cooling.

Restraint classification can apply to various types of building construction (e.g., steel, concrete, and timber); however, it is most commonly encountered for steel construction. Several organizations have conducted furnace testing to specifically address the ASTM E119 restraint classifications. AISC and AISI funded furnace testing of steel floor assemblies, which found that restraint of the furnace testing frame provided no significant fire resistance benefit (Carter and Alfawakhiri 2013, 2017). NIST performed furnace testing of composite floor assemblies with reinforced concrete slabs and steel trusses and found that an *unrestrained* assembly achieved greater fire resistance when compared to an equivalent *restrained* assembly (Gross et al. 2005). These test results demonstrate that the effect of restraint varies among different structural systems and restraint conditions, and cannot be easily simplified, especially in a binary fashion (i.e., "restrained" vs. "unrestrained").

For standard fire tests of floors and roofs, design gravity loads are applied "to simulate a maximum load condition," which is determined as "the maximum load condition allowed under nationally recognized structural design criteria" (ASTM 2016). For tests of steel columns, loading is not required, and it is rarely used during a standard fire test. Neither of these load conditions (full dead and live load or no load) reflect the design conditions given by the load combinations for extraordinary events (see Section 10.1). Accordingly, it is unlikely that a structure will experience design-level gravity loads and a structurally significant fire event at the same time (Ellingwood and Dusenberry 2005).

Assemblies that are qualified through standard fire testing are published in directories such the *UL Fire Resistance Directories* (UL 2015), which contain hourly ratings for beams, floors, roofs, columns, walls, and partitions. The application of standard fire ratings in building construction over the last century has greatly improved the performance of buildings subject to structurally significant fires. However, because of its comparative basis and the lack of structural performance information, standard fire testing is not intended to evaluate the performance of an assembly under actual fire conditions or within a structural system.

2.2.2 Calculation Methods

Standard fire tests can be very costly to conduct. When members or assemblies and their passive fire protection are similar to those already tested, certain calculation methods may be used to determine the fire resistance rating. These calculations, which are referred to as equivalence methods, generally interpolate between established test data. Extrapolation from published fire ratings would not be appropriate given the empirical nature of standard fire testing.

ASCE/SEI/SFPE 29-05 provides specific calculations methods to determine the fire resistance rating of certain assembly types (ASCE/SEI/SFPE 2005). Calculation methods are provided for structural members and fire barrier assemblies made from structural steel, plain concrete, reinforced concrete, timber and wood, concrete masonry, and clay masonry.

2.3 STRUCTURAL FIRE ENGINEERING DESIGN (SFED)

SFED is based on measures of *strength* given the consideration of potential structural design fires. In accordance with the fundamental philosophy used to design structures, SFED explicitly evaluates demand and capacity of structural systems under fire exposure. Within this framework, the demand on a structural system under fire exposure can be reduced by means of rationally allocated structural insulation (e.g., SFRM), control of fuel loads, and/or other fire exposure mitigation techniques. Also, the capacity of a structural system to endure fire exposure can be increased by means of specific member sizing, connection detailing, and/or other measures to enhance

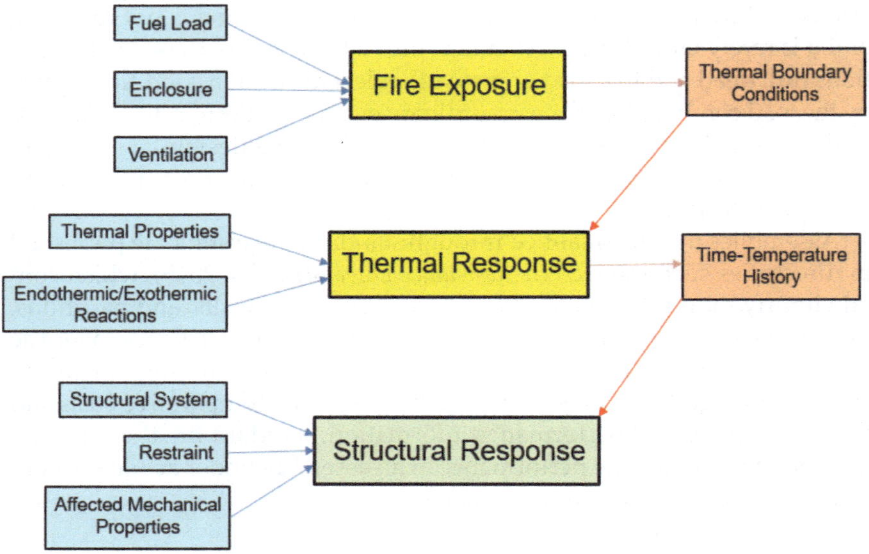

Figure 2-4. Generalized design process for SFED.
Source: Kevin LaMalva, reproduced with permission.

structural robustness. Figure 2-4 illustrates the generalized design process for SFED, which is deterministic and sequentially coupled.

SFED methods are often applied to a structural design that results from adherence to requirements for primary design loads (e.g., dead, live, wind, seismic, and snow). However, contemplation of fire performance may be part of the initial structural design as well. SFED does not rely upon standard furnace testing or other qualification methods. Accordingly, SFED is appropriate for conditions where SFRD would not properly address the design objectives of stakeholders. For example, SFED is appropriate for cases where structural performance under fire exposure needs to be quantified to properly assess the risks. Also, this approach may be required if the owner seeks to meet specific performance objectives, such as enhanced property protection (e.g., minimization of physical damage and business interruption). Lastly, building code variances for structural fire protection require a SFED approach to demonstrate the adequacy of an alternative design. In terms of design flexibility, SFED allows the designer to control a greater number of design variables as compared to SFRD (Figure 2-5).

2.3.1 Professional Responsibility

Unlike SFRD, in which the architect typically serves as the responsible party for satisfying code requirements for structural fire protection, SFED

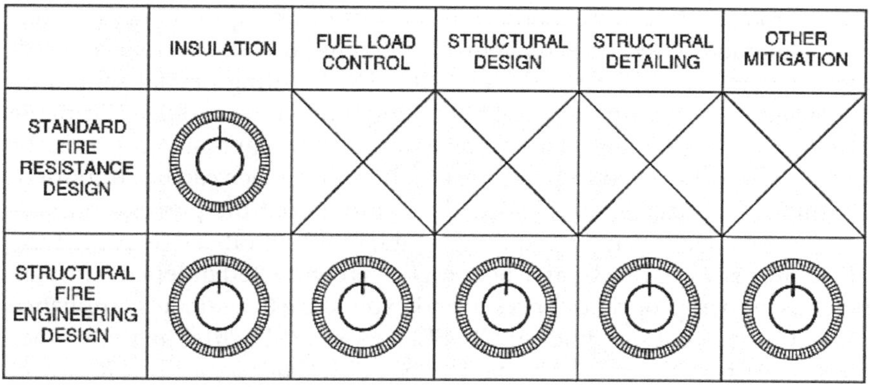

Figure 2-5. Controllable design variables.
Source: LaMalva (2017a), reproduced with permission.

usually requires a team consisting of structural engineers, fire protection engineers, and other design professionals as needed. Also, the owner, building authority, and possibly third-party peer reviewers need to be included as part of the design process. Accordingly, the scope of tasks and professional responsibilities of each stakeholder should be clearly defined in project documentation or contracts.

For SFRD, the architect usually selects qualified fire resistant assemblies from available listings. This process typically does not require appreciable engineering participation or rigor. Hence, fire protection engineers and/or structural engineers are typically involved on a case-by-case basis for unusual or complex situations only. For SFED, the fire protection engineer determines structural design fire scenarios, evaluates the mechanical integrity of proposed fire protection methods, and determines the thermal environment and response (temperature histories) of structural systems. These tasks are important because the selection of proper structural design fires and quantification of their heating to the structural system may be critical to the structural response. The structural engineer may assist the fire protection engineer in determining which structural systems should be evaluated for the structural design fire scenarios. The structural engineer's primary responsibility is to evaluate the response of the affected structural system to fire effects, such as effects of thermal expansion and reduced stiffness and strength. Structural engineers with expertise/experience in SFED are commonly referred to as *structural fire engineers*. The architect would also be involved with this process, as proposed solutions may affect other aspects of the building design or interior spaces.

The owner, building official, and any third-party peer reviewers also need to be involved throughout the process. The owner should understand the need for SFED, how the process works, and the costs associated with the approach. The building official is responsible for approving this alternative method based on achievement of the required performance objectives per ASCE/SEI 7 (ASCE/SEI 2016). The building official should be contacted prior to initiating a SFED to determine if their office is capable of supporting such alternative means or if they have any concerns or requirements that need to be addressed. If the approach can be supported, the building official should be involved throughout the process. A third-party peer review may be required by the building official as part of the SFED process. The third-party peer review is conducted independently by persons with appropriate expertise and experience to evaluate compliance of the proposed design with ASCE/SEI 7 (ASCE/SEI 2016) and any other requirements of the building official. The SFPE *Guideline for Peer Review in the Fire Protection Design Process* (SFPE 2009) provides guidance on the peer review process.

The effects of fire exposure on structural systems should be evaluated as an extraordinary event as defined in Section 2.5 of ASCE/SEI 7 (ASCE/SEI 2016). If fire performance was not contemplated during the initial design, the structural engineer may need to propose modifications to the structural system to achieve satisfactory performance for design fire conditions.

REFERENCES

ANSI/UL (American National Standards Institute/Underwriters Laboratories). 2011. *Standard for fire tests of building construction and materials.* ANSI/UL 263. Northbrook, IL: ANSI/UL.

ASCE/SEI (Structural Engineering Institute). 2016. *Minimum design loads and associated criteria for buildings and other structures.* ASCE/SEI 7. Reston, VA: ASCE.

ASCE/SEI/SFPE (ASCE/Structural Engineering Institute/Society of Fire Protection Engineers). 2005. *Standard calculation methods for structural fire protection.* ASCE/SEI/SFPE 29-05. Reston, VA: ASCE.

ASTM International. 1971. *Standard test methods for fire tests of building construction and materials.* E119-71. West Conshohocken, PA: ASTM International.

ASTM. 1973. *Standard test methods for fire tests of building construction and materials.* E119-73. West Conshohocken, PA: ASTM.

ASTM. 2010. *Standard test methods for determining effects of large hydrocarbon pool fires on structural members and assemblies.* E1529-10. West Conshohocken, PA: ASTM.

ASTM. 2016. *Standard test methods for fire tests of building construction and materials.* E119-16a. West Conshohocken, PA: ASTM.

BS (British Standard). 1987. *Fire tests on building materials and structures*. BS 476-20:1987. London: British Standard International.

Carter, C. J., and F. Alfawakhiri. 2013. "Restrained or unrestrained?" *Mod. Steel Constr.* Accessed September 2013. https://www.aisc.org/modern-steel/archives/2013/september/.

Carter, C. J. and F. Alfawakhiri. 2017. "Comprehensive coverage." *Mod. Steel Const.* Accessed September 2013. https://www.aisc.org/globalassets/modern-steel/archives/2017/01/sw.pdf.

CEN (European Committee for Standardization). 1999. *Fire resistance tests*. EN1363. Brussels: CEN.

Ellingwood, B., and D. Dusenberry. 2005. "Building design for abnormal loads and progressive collapse." *Comput.-Aided Civ. Infrastruct. Eng.* 20(3), 194–205.

Goode, M. (ed.). 2004. "Fire protection of structural steel in high-rise buildings." *NIST GCR 04-872*. Gaithersburg, MD: NIST.

Gross, J., F. Harvey, M. Isydorek, J. Mammoser, and J. Treadway. 2005. "Federal building and fire safety investigation of the World Trade Center disaster: Fire resistance tests of floor truss systems." *NIST NCSTAR (National Construction Safety Team Act Reports) 1-6B*. Gaithersburg, MD: NIST.

ICC (International Code Council). 2012. *2012 International building code*. Falls Church, VA: International Code Council.

ISO. 1999. *ISO 834-1:1999 Fire resistance tests—Elements of building construction—Part 1: General requirements*. Geneva: ISO.

LaMalva, K. 2017a. "Fire design." In *Architect's guidebooks to structures: special structural topics*, P. McMullin, J. S. Price, and S. Simchuk (eds.). New York: Routledge.

LaMalva, K. 2017b. "Reexamination of 'restrained vs. unrestrained.'" *FPE Extra*, 20, Bethesda, MD: Society of Fire Protection Engineers.

LaMalva, K. 2018. "Developments in structural fire protection design—A U.S. perspective." *Struct. Eng.* (96)1, 25–29.

Lamont, S. 2001. "The behavior of multi-storey composite steel framed structures in response to compartment fires." Ph.D. thesis, College of Science and Engineering, School of Engineering and Electronics, Univ. of Edinburgh, Edinburgh, Scotland.

Nelson, H. 2003. *History of fire protection engineering*. Quincy, MA: National Fire Protection Association.

NFPA (National Fire Protection Association). 1928. "Fire test of brick joisted buildings." *NFPA Quart.* 22(1).

SFPE (Society of Fire Protection Engineers). 2009. *Guidelines for peer review in the fire protection design process*. Bethesda, MD: Society of Fire Protection Engineers.

UBC (Uniform Building Code). 1928. *1927 Uniform Building Code*. Long Beach, CA: International Conference of Building Officials. Accessed

June 20, 2018. http://shop.iccsafe.org/1927-uniform-building-code-download-1.html.

UL (Underwriters Laboratories). 2011. *Standard for rapid rise fire tests of protection materials for structural steel.* UL1709. Northbrook, IL: Underwriters Laboratories.

UL (Underwriters Laboratories). 2015. *UL fire resistance directory,* Vol. 1. Northbrook, IL: Underwriters Laboratories.

CHAPTER 3
PERFORMANCE OBJECTIVES

Structural systems with members or assemblies that are qualified through standard fire testing typically use applied insulation to delay heating and to achieve the fire rating mandated by building codes. However, structural systems designed with SFRD are not designed to resist fire effects, such as reduced stiffness and strength in members, and loads induced by thermal restraint (if occurring). SFED provides a rational basis for determining if a structural system can meet the intended performance objectives under fire exposure. SFED compares the structural response to fire effects (estimated by simple calculations and/or advanced simulation) against explicit performance objectives to evaluate the adequacy of the proposed structural design and applied insulation.

Specific performance objectives are not provided in all building codes. However, if Appendix E in ASCE/SEI 7 (ASCE/SEI 2016) is adopted, it can provide minimum life safety performance objectives for structural systems under fire exposure, as well as guidance for other desired performance objectives.

3.1 MINIMUM PERFORMANCE OBJECTIVES

ASCE/SEI 7, Appendix E, provides minimum performance objectives that pertain to occupant life safety. Specifically, the performance of the structural system under structural design fires must allow for building occupants to travel safely to refuge areas within the building or exit the building to a public way (e.g., a public street). Accordingly, structural support of building egress routes must be maintained for the period of time necessary to ensure safe travel to refuge areas within the building or safe exit from the building to a public way. Also, structural support of building refuge areas must be maintained throughout the heating and cooling of the structure (ASCE/SEI 2016).

Building codes limit egress travel distances to exits (e.g., stairways), but they generally do not limit the total evacuation time. As the vertical remoteness of occupants from the point of discharge to a public way is increased, the time required to evacuate the building will increase. Unlike SFRD, SFED explicitly evaluates the consequences of increased occupant evacuation times and the reliance on building refuge areas to meet other code requirements.

To demonstrate the adequacy of occupant egress routes, the designer should use an *ASET vs. RSET* analysis. The available safe egress time (ASET) should be determined by analyzing the endurance of the structural system to fire exposure. The required safe egress time (RSET) should be determined by analyzing the time it would take occupants to travel safely to refuge areas within the building or to exit the building to a public way. RSET is commonly estimated using evacuation models that can range in complexity from simple hand calculations to advanced computer-based models that estimate human behaviors. The *SFPE Guide to Human Behavior in Fire* (SFPE 2003) provides guidance on how to calculate RSET. In an ASET versus RSET analysis, the ASET value should be greater than the RSET value. Alternatively, the designer may demonstrate that structural endurance is provided through full fire burnout.

In addition to the consideration of occupant egress routes, the structural system that supports refuge areas must remain stable during and after an uncontrolled fire exposure. The evaluation of RSET and identification of building refuge areas is best performed by a fire protection engineer.

3.2 DISCRETIONARY PERFORMANCE OBJECTIVES

A greater level of structural performance may be desired for a given project than that provided by adherence to the minimum performance objectives only. For these cases, ASCE/SEI 7 Appendix E provides guidance for identifying discretionary (project-specific) performance objectives. For property protection, Appendix E defines tolerances for structural damage based on the risk category of the building (ASCE/SEI 2016). For instance, essential facilities such as hospitals or emergency response facilities may need to remain functional during and after a structural design fire event.

The need for and extent of discretionary performance objectives shall be determined by the project stakeholders, most notably the owner and building authority. At the outset of the SFED, the discretionary performance objectives that have been agreed upon by all the stakeholders should be documented within a design brief document in accordance with the SFPE *Engineering Guide to Performance-Based Fire Protection* (SFPE 2007).

Discretionary performance objectives may address issues such as recovery of function, property protection, business continuity, environmental protection, adequate structural support of fire resistance-rated assemblies to limit fire and smoke spread, and structural support for first responder operations. Satisfaction of the mandatory performance objectives may implicitly enhance these measures of performance.

REFERENCES

ASCE/SEI (Structural Engineering Institute). 2016. *Minimum design loads and associated criteria for buildings and other structures.* ASCE/SEI 7. Reston, VA: ASCE.

SFPE (Society of Fire Protection Engineers). 2003. *Engineering guide to human behavior in fire.* Bethesda, MD: Society of Fire Protection Engineers.

SFPE. 2007. *Engineering guide to performance-based fire protection,* 2nd Ed. Bethesda, MD: Society of Fire Protection Engineers.

PART 2

THERMAL RESPONSE OF STRUCTURES UNDER FIRE

CHAPTER 4
THERMAL BOUNDARY CONDITIONS

ASCE/SEI 7 (ASCE/SEI 2016) Section E.5 provides requirements for calculating structural component temperature histories due to fire exposure, which serve as input for subsequent structural response analyses. Determination of structural component temperature histories involves the following requisite tasks: quantification of the fuel load, evaluation of structural design fires, and determination of the structure's thermal response. This chapter examines the first two requisite tasks that characterize the thermal boundary conditions on the structural system under uncontrolled fire conditions.

4.1 FUEL LOAD

Fuel load is commonly expressed as a fuel load density or fuel load per unit floor area. The NFPA 557 standard establishes a basis for selecting fuel load density and distribution for SFED (NFPA 2016). Although the fuel load density concept implies a uniform distribution of combustibles in compartments, the actual distribution of combustibles may need to be addressed in analyses of structural design fire scenarios.

NFPA 557 (NFPA 2016) has two primary methods for estimating fuel load density: the occupancy-based method and the survey-based method. The occupancy-based method only requires information contained in the standard to compute a fuel load density. The survey-based method serves as an alternative approach and requires data specific to the project. Generally, a survey-based approach would only be used in cases where justification of a lower fuel load density (as compared to that calculated using the occupancy-based method) is warranted by special circumstances. Otherwise, there would be limited justification for the effort involved.

The occupancy-based method requires the determination of a specific risk factor parameter that represents the likelihood of high heating to the structural system. This risk factor is based on the level of protection provided by structural insulation, fire sprinkler systems, and smoke/heat detection systems. The standard provides average and standard deviation values for fuel load densities based on statistical data from published fuel load surveys. For instance, business occupancies are specified to have average and standard deviation values of 600 MJ/m^2 and 500 MJ/m^2, respectively (NFPA 2016). Eurocode 1 (CEN 2001) provides a similar framework for determination of fuel load density, which is a function of the nominal fuel load and specific risk factors per Annex E. Notably, there is a risk factor that accounts for the presence or absence of active fire protection measures (including manual suppression), which is similar to NFPA 557.

The survey-based method in NFPA 557 involves the manual accounting of anticipated fuel masses and their respective combustion properties. Various combustible materials release varying level of thermal energy when they burn. For instance, many hydrocarbon-based materials (e.g., plastics) have approximately twice the heat of combustion of cellulosic materials (e.g., wood). Specifically, burning one kilogram of a plastic can liberate twice the energy as burning an equal mass of wood. Also, certain plastics can liberate more than twice the energy as compared to wood. To account for this, the standard requires that individual fuel masses be converted to wood equivalent mass based on their potential energies.

If a survey-based approach is adopted for quantification of the design fuel load, the designer should provide specific data. In all cases, the design fuel load should be memorialized within a design brief document prior to conducting analyses. If the building changes occupancy type in the future, the new fuel load levels should be reevaluated. Also, if the fuel loads of a building are anticipated to change over time (e.g., arenas with multiple uses), the worst-case fuel loads should be considered.

4.2 FIRE EXPOSURE CONDITIONS

Except for large open spaces, fire in enclosures may be characterized in three phases. The first phase is fire growth, when a fire grows in size and heat release rate from a small incipient fire. If there are no actions taken to suppress the fire, it will eventually grow to a maximum size (fully developed phase), which is a function of the amount of fuel present, the amount of air available through ventilation openings, and the geometry of the compartment. As the fuel is consumed, the fire will decrease in size during the decay phase. These stages of fire development are illustrated in Figure 4-1.

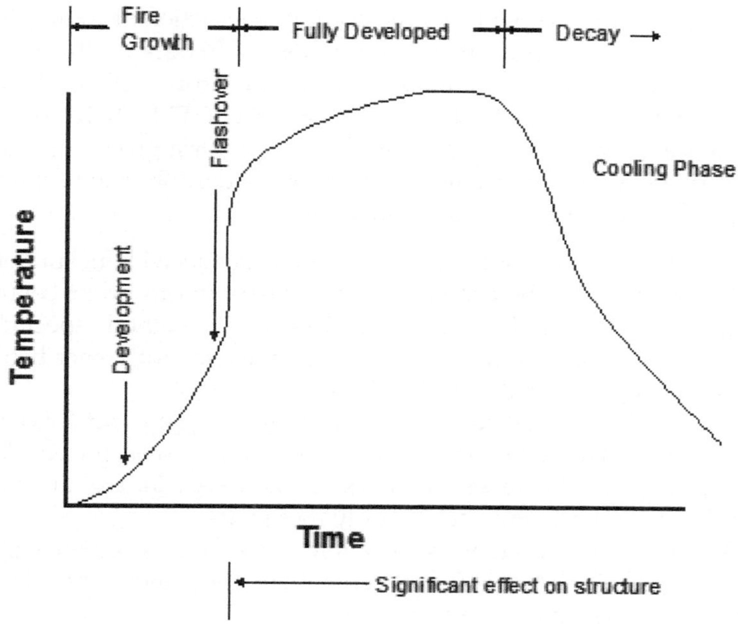

Figure 4-1. Phases of fire development.
Source: SFPE (2004), reprinted with permission.

The definition of a structural design fire stipulates that there are no actions taken to suppress the fire.

The fire scenario that is most commonly considered in SFED is a fully developed enclosure fire. In most fires, there can be variability in the fire size depending on which items are ignited first and how the fire grows and spreads during the growth phase of a fire. Focusing on fully developed fires eliminates much of this variability because a fully developed (post-flashover) fire is less sensitive to which items are ignited first and how the fire grows. For SFED cases involving analysis of compartment boundaries, neglecting the heating to the structure during the growth stage of a fire is reasonable because the heating of the structure during the fully developed stage is much greater than the heating during the growth stage. Localized fires are also commonly considered, where a fire may affect nearby members or assemblies, but does not reach flashover conditions.

Design equations have been developed for fully developed enclosure fires, as well as for localized fires (SFPE 2011b). Alternatively, computational analyses can be conducted using zone or field models to develop a more comprehensive understanding of the effects of natural fire exposure that are beyond the scope of existing design equations. However, zone and field models may not provide a more accurate prediction of compartment time–temperature histories than simpler models do because the assumptions embedded within these models may not be applicable to the post-flashover fire environment. The SFPE G.06 guide (SFPE 2011a) provides guidance for selecting an appropriate fire model for a given application. Aside from field models, predictive methods (as described in Section 4.3) are often subject to the following limitations:

1. The methods are only applicable to compartments with fuel uniformly distributed over their interior. Sparse distributions or concentrated fuel packages should be considered using the methods specific to a localized fire exposure. Figure 4-2 illustrates the difference between an enclosure fire and a localized fire.
2. Most predictive methods are only applicable to compartments having vents in walls. Ceiling and floor vents require a special formulation (e.g., use of a field model as described in Section 4.3.3), as would underground compartments having only roof vents.
3. Only natural ventilation is considered as would occur through the wall vents. The effect of forced ventilations and wind and stack-effect flows in tall buildings are not included.

Based upon the fuel load, ignition(s), and the arrangement of compartments and openings, a fire exposure may be broadly defined as one of the following types: enclosure fire, localized fire, spreading/traveling fire, post-earthquake fire, and exterior fire exposure. The following sections examine these fire exposure types in detail.

Figure 4-2. Enclosure fire (left) versus localized fire (right) exposure. Source: LaMalva (2017), reprinted with permission.

4.2.1 Enclosure Fires

A fire that is affected by the compartment(s) in which it resides is referred to as an enclosure fire. In this case, a hot upper gas layer forms and is able to radiate heat back to the fire and surrounding fuel packages. Also, the compartment boundaries may re-radiate inward as well. These conditions may lead to flashover if the fire is able to grow uninterrupted with an adequate air supply, at which point the fire is considered to be *fully developed*. In these instances, an enclosure fire model would be appropriate. For a given fire compartment, the post-flashover intensity and duration should be estimated as a function of the following parameters:

- Total area of enclosure boundaries;
- Total area of ventilation openings;
- Height of ventilation openings (weighted average);
- Density of enclosure boundaries;
- Thermal conductivity of enclosure boundaries; and
- Specific heat of enclosure boundaries.

Enclosure fire models provide the gas temperature history of the compartment. The SFPE S.01 standard provides methodologies to predict fire exposure conditions from a fully developed fire (SFPE 2011b). As a basis for the design equations in SFPE S.01, SFPE collected and synthesized data from 139 fully developed fire tests. This database is primarily composed of cellulosic-type fires; however, about 10% are chemical in nature and about 20% involve mixed Class A materials. The design equations in SFPE S.01 are very conservative because they produce results with a very low probability of exceedance in terms of the fire exposure intensity as compared to the database. However, the resulting time–temperature histories have a relative low correlation with the actual fire curves in the database (Hughes Associates 2006).

As an alternative to SFPE S.01, the designer may use design equations from Annex A of Eurocode 1 (CEN 2001) to derive temperature histories of enclosure fires. Annex A provides equations for parametric fire curves that require the input of the relevant parameters listed previously. These design equations are not as conservative as those in SFPE S.01; however, they have a higher correlation coefficient when compared to actual enclosure fires within the SFPE database (Hughes Associates 2006). The designer should be aware that the Eurocode Annex A equations are only valid for enclosure areas up to 100 m^2 (approximately 32 ft × 32 ft) and that many building spaces exceed this limit (Jonsdottir and Rein 2009). However, it has been argued that applying these equations to areas exceeding 100 m^2 is conservative given that smaller enclosures would contain more heat (Kirby et al. 1999). Alternatively, for large compartments, the designer may choose to analyze a more realistic case, in which a traveling fire occurs without flashover, as discussed in Section 4.2.4.

In addition to SFPE S.01 and the Eurocode, several methods have been published in the literature for estimating the time–temperature history of fully developed compartment fires (SFPE 2004). Most predictive methods assume a one-zone compartment, in which the gas temperature is modeled as uniform throughout the compartment at any given time. Most predictive methods available in the literature have their basis in fires involving wood as the fuel. As discussed in Section 4.1, various combustible materials release varying levels of thermal energy when they burn. The use of methods that assume wood as the fuel should be reasonable for most structural design fire scenarios in buildings as long as the fuel load is appropriately converted to a wood equivalent mass. Design or analysis situations in which the fuels are not predominantly cellulosic and the burning is not expected to be ventilation-limited (e.g., enclosure fire is controlled by air supply) may require special attention.

4.2.2 Localized Fires

Fires may produce localized heating of the structure. This is typically the case in relatively large compartments or when the fuel is concentrated within a certain region in the compartment. In these instances, a localized fire model may be appropriate. For a given nonenclosed area, the fire exposure intensity and duration should be estimated based on the fuel load density and distribution. SFPE S.01 provides methodologies to predict fire exposure conditions from a localized fire involving a concentrated fuel load that is not affected by the compartment (SFPE 2011b). Localized fire models provide the heat flux or temperature acting on the fire-exposed surfaces assuming that the fire does not reach flashover. In all cases, flame height calculations should be conducted to determine if flame impingement would occur. Also, the duration of the fire should be calculated.

SFPE S.01 provides appropriate equations for localized fires beneath a ceiling, fires adjacent to a flat wall, and fires adjacent to a corner (SFPE 2011b). The standard also provides the surface heat flux and burning duration based on published data from fire tests. Lastly, the standard provides guidance on selecting appropriate localized structural fire scenarios. The SFPE *Engineering Guide to Fire Exposures to Structural Elements* (SFPE 2004) provides a comparison among alternate methods for assessing localized fire exposures, and Lattimer (2016) gives additional background information regarding heat fluxes from localized fires.

4.2.3 Exterior Fire Exposure

Fires that occur outside a given building may lead to ignition and subsequent fire exposure within the subject building, perhaps on multiple floors simultaneously. Also, buildings in close proximity may mutually increase the risk of a large fire exposure. Lastly, wildland–urban interface conditions may increase the risk of exterior fire exposure, depending upon the specific protection provided (e.g., fire shutters). Radiant heat can occur at distances substantially greater than those at which flame impingement and convective heat transfer usually constitute a hazard. Therefore, ignition as a result of radiation heat transfer usually governs the hazard from an exterior fire exposure.

Buildings with combustible facades have an increased risk of exterior fire exposure. In these cases, fires within the building may spread to the exterior of the building through openings (e.g., broken windows). For such cases, Annex B of Eurocode 3 (CEN 2010) provides calculation methods to estimate the heating from exterior flaming out of building openings. These methods include scenarios in which the exterior structure is engulfed in fire or at a distance and exposed to radiation. In addition to Eurocode 3, Law and O'Brien (1983) provide a detailed method for calculating flame size, height, horizontal projection distance, and temperature of exterior flames from building openings with and without wind conditions.

Structural design fire scenarios primarily involve those that may originate within the subject building. However, the building authority may require that the designer consider the risk associated with exterior fire exposures from an adjacent building or hazard. The NFPA 80A standard provides guidance concerning the hazard of an exterior fire exposure from a building in close proximity (NFPA 2012). This standard addresses separation distances between buildings to limit exterior fire spread based on exterior openings and other construction features. Recommended building separation distances are primarily based on a maximum tolerable incident radiation at the exterior face of an exposed building of 12.5 kW/m^2 for ordinary combustibles such as wood. This standard is also intended to provide a reasonable level of protection for combustibles within and on the exterior of a building exposed to an exterior building fire. Also, the SFPE *Engineering Guide to Assessing Flame Radiation to External Targets from Pool Fires* (SFPE 1999) provides guidance for calculating exposure conditions to objects or surfaces that are heated by flame radiation only.

Exposure to radiant heat from a nearby building may result from any of the following: radiant heat passing through windows or other openings in the exterior of a burning building, flames extending from the windows of a burning building, and flames extending from the burning exterior of a building. NFPA 80A recommends that the height of the exposing fire be a function of the height of the number of stories involved in the fire, considering such factors as the building construction, closure of vertical openings, and fire resistance of floors (NFPA 2012). Furthermore, the width of the exposing fire should be considered to be the length of the exposing wall between interior fire separations (e.g., fire barrier walls) or between exterior end walls if no fire separations exist. Lastly, the percentage of openings

Table 4-1. Average Heights of Flames Penetrating Roofs.

Number of stories burning	Flame height above roof (in stories)
1	1.4
2	1.8
3	2.2
4	2.6
5	2.9
6	3.1

Source: Nelson (1968).

in the exposing wall area should be regarded as the percentage of the exposing wall made up of doors, windows, or other openings within the assumed height and width of the exposing fire.

The exposure hazard of a given building in close proximity to a burning building is a combination of radiation emitted through the windows and that produced from the flames that project out the windows and up the front of the burning building. Because radiation from the flames, as well as from the interior room walls of the burning building, are involved, the flame emissivity could be of great importance. It is generally assumed for radiation calculation purposes that the emissivity is 1.0, which represents a *blackbody* radiator. Flames of a burning building may also compromise the integrity of a roof and extend higher than the building height.

Table 4-1 correlates the number of stories involved in fire and the resulting flame height above the roof line if the fire is not contained by the roof (Nelson 1968). In the event of moderate wind, flames can be expected to extend horizontally to a distance equal to that which they might otherwise extend upward.

The extent of exterior openings and the flammability of the building's exterior are important factors to consider as it pertains to the risk of a fire event from an exterior fire exposure. Generally, greater separation distances or fewer openings are needed for materials with greater propensity to ignite. Materials offering greater resistance to ignition can be separated at lesser distances or more openings can be permitted. ASTM E136 (ASTM 2011) and ASTM E2652 (ASTM 2016) provide test methods to determine the combustion characteristics of building materials.

4.2.4 Traveling Fires

For a given fire exposure, the fire may move from one area to another as contiguous fuel packages are ignited and/or as compartment boundaries are compromised. Also, traveling fires can occur when the fire is fully developed if there is insufficient ventilation available to support combustion of all items in the compartment. In such cases, the fire will burn preferentially in the vicinity of ventilation openings, such as doors or windows. Items that are farther away from the ventilation openings will only burn once the items closer to the ventilation openings are consumed. Currently, the SFPE S.01 standard does not provide guidance for traveling fires (SFPE 2011b), and there is a limited amount of research available on the subject. Hence, a designer should exercise conservatism when analyzing traveling fire scenarios.

Inspection of historic fires in large, open-plan compartments reveals that they do not burn simultaneously throughout the whole enclosure. Instead, these fires tend to move across floor plans as the fire spreads, burning over a limited area at any one time. Many large fires, such as those in the World Trade Center towers 1, 2, and 7 in 2001; the Windsor Tower in 2005; and the Faculty of Architecture Building at TU Delft in 2008, were all observed to travel across floor plans and vertically between floors rather than burn uniformly for their duration. Similar observations were made of the First Interstate Bank fire in 1988 and the One Meridian Plaza fire in 1991. Traveling fires have also been observed experimentally in compartments with nonuniform ventilation (Stern-Gottfried 2012a and 2012b).

Traveling fires do not uniformly heat the structure of a given compartment, which may have a considerable impact on the structural system behavior. Stern-Gottfried (2012) developed a design methodology using traveling fires to produce more realistic fire scenarios in large, open-plan compartments. However, this approach has not been validated to large-scale test data and may be viewed as overly complex for practical application. Hence, this method has not been incorporated into key industry standards, such as SFPE S.01. For the open-plan fire scenarios considered, Stern-Gottfried found that the peak temperature of a given structural system occurred at the final burning location of the fire. Hence, it was reasoned that designers could identify particular areas of a given structure that are most vulnerable to the effects of a traveling fire, and then it can be conservatively assumed that the fire reaches this location last, thereby producing the most onerous fire environment for that part of the structure. Rackauskaite et al. (2015) and Rackauskaite et al. (2017) further explored and refined the methodology developed by Stern-Gottfried (2012a and 2012b).

The extent of a structural design fire is typically considered to be limited to areas that are bounded by fire-rated construction and/or exterior walls of a building. It is well known that standard furnace testing does not provide an adequate measure of in-situ structural performance. However, standard furnace testing of fire barriers does provide a reasonable measure of in-situ flame resistance performance. Granted, the designer would need to exercise caution in cases where the structural design fire is significantly more severe and/or has a longer duration as compared to the standard furnace exposure and the associated fire resistance rating. Accordingly, the NFPA 557 (NFPA 2016) and SFPE S.01 (SFPE 2011b) standards specify the extent of a structural design fire as either the entire floor or that portion of the floor that is bounded by exterior walls and/or by fire-rated boundaries that are capable of containing a fire for the entire duration through burnout.

If a fire starts in a compartment with walls and floors of fire resistance insufficient to contain the fire, it may be necessary to consider two fire spread scenarios. First, it could be assumed that the compartment's boundary fails under fire exposure at the beginning of the fire, leading to a larger room of origin. Second, it could be assumed that the compartment's boundary is sufficient to contain the fire for a significant time period but later fails and is no longer effective. As a consequence, the fire will spread beyond the compartment of origin. It may be necessary to consider both possibilities because it is generally difficult to know which case would govern the structural design (CIB 2001). Fitzgerald and Meacham (2017) discuss concepts for assessing the performance of fire barriers for resisting fire spread, which may be applied to analysis of traveling fires. The risk associated with traveling fires and the necessity to consider such scenarios should be determined by a fire protection engineer or other qualified professional.

4.2.5 Post-Hazard Event Fires

Hazard events such as an earthquake or flood may result in one or a combination of enclosure, localized, and/or traveling fires within a building. Post-hazard event fires may have compounding factors such as dispersed flammable contents and electrical malfunctions. Also, structural damage caused by a severe seismic event may be exacerbated by fire exposure (e.g., exposed concrete reinforcement). ASCE/SEI 7 Appendix E does not pertain to post-hazard event fires (ASCE/SEI 2016). Accordingly, designers seeking to analyze such cases should exercise caution for there is limited relevant guidance and research. Notable testing and research on structural behavior under post-hazard event fires include Wang et al. (2016) and Khorasani et al. (2016). In any case, the risk associated with post-hazard event fires and the necessity to consider such scenarios should be determined by a fire protection engineer or other qualified professional. At a minimum, the damaged state of the structural system should be included in analyses of post-hazard event fires. This section provides some commentary on the matter.

History has demonstrated that earthquakes significantly raise the risk of fire exposure within buildings. For instance, nearly all major earthquakes in California during modern times have caused fires in both residential and commercial facilities (Mohammadi et al. 1992). For modern buildings, significant structural damage is expected to occur at high earthquake intensity levels. However, lesser levels of structural damage may have implications on fire resiliency. Significant cracking or movement may allow fire spread through walls and slabs, as well as dislodgement of structural insulation from structural members and elements. Damage to stairs may hinder evacuation from the building. Loss of exterior glazing from an earthquake may change the ventilation factors for a fire compartment and allow the development of large fuel-controlled fires (Taylor 2003), as well as increase the risk of vertical flame spread to floors above the fire.

Although there are many possible causes of fire following earthquakes, some of the most common appear to be fires associated with natural gas leaks, fires caused by electrical system malfunctions, and fires caused by hazardous chemical interactions or involving flammable liquids (NIST 1995). In the United States, natural gas (methane) is a factor from 15% to 50% of the time, and electrical ignitions typically account for 40% or more of post-earthquake fires (FM Global 2001). Water heaters and other gas appliances can topple or slide when subjected to ground motion due to earthquakes. Movement of gas appliances can break gas service connections, resulting in the release and possible ignition of natural gas in a building (NIST 1995).

Fires following earthquakes occur not only in the immediate aftermath of the earthquake, but the incidence of fires generally continues at a higher-than-normal rate for several days. This can be attributed at least in part to fires resulting from utility restoration and cleanup efforts (NIST 1995). Typically, 50% to 75% of earthquake-initiated fires start immediately after the event (FM Global 2001). Electric power often fails in large earthquakes because of automatic system trips, as well as damage to the system. However, the power failure usually takes several seconds, during which time power could be a source of multiple ignitions. Certain electric appliances (e.g., those with heating elements) can still cause fires even after power is cut (Scawthorn 2008). When power is restored, short circuits that occurred because of the earthquake become energized and may ignite fires. Equipment or cooling system damage may result in overheating and fire in equipment that remains operating (FM Global 2001).

Damage to nonstructural partitions is a function of the racking loads imposed by the deflections of the building because of an earthquake. Damage is therefore more likely in flexible framed buildings unless the partitions are separated from the structure to allow differential movement. The performance will depend on the flexibility of the building, as well as on the details of the enclosure panels, such as aspect ratio, presence of doors and penetrations, and degree of edge restraint from the main structure. Successful structural separation depends on properly designed and detailed fire resistant seismic joints. Most modern seismic building codes contain provisions for seismic restraint of building fittings and equipment and appropriate separation of nonstructural elements that are vulnerable to earthquake damage (Taylor 2003).

This discussion provides an overview of conditions that may result in post-hazard event fires. The risk associated with post-hazard event fires and the necessity to consider such scenarios should be determined by a fire protection engineer or other qualified professional.

4.3 FIRE EXPOSURE CALCULATION METHODS

Enclosure fires and localized fires can be readily modeled using calculation methods defined in the SFPE S.01 standard (SFPE 2011b). For enclosure fires, design equations have been developed to determine the uniform gas temperature as a function of time based on factors such as the fuel load density, ventilation factor, and compartment characteristics. For localized fires, specified heat flux values are provided for smoke or flame impingement. Other equivalent calculation methods can be used if validated against actual fire experiments.

In special circumstances, it may be necessary to perform a computational fire analysis using a zone or field model. It should be noted that fire modeling using zone and field models generally requires the expertise of a fire protection engineer or other qualified professional. Also, the refined spatial fidelity of zone or field models may not correspond to improved accuracy when compared to algebraic correlations. Prior to using zone or field models, the designer should ensure that the assumptions embedded within the model correspond to the compartment and fire environment being modeled. Most fire models created in practice simulate or estimate the effects of fire and not the phenomena of combustion and flame spread. For instance, a designer may prescribe a heat release rate history of a hypothetical fire source to estimate the resulting temperature histories throughout the building spaces. Modeling combustion and flame spread is possible, but it should be conducted with caution and with significant validation to actual fire experiments.

4.3.1 Algebraic Correlations

Enclosure fires are generally represented by a single temperature–time curve based on post-flashover conditions that are ventilation-limited. SFPE S.01 provides two calculation methods for evaluating gas temperature of post-flashover fires (SFPE 2011b): (1) the constant compartment temperature method, which specifies a constant gas temperature of 1200°C for the duration of burning, and (2) the Tanaka (refined) method, which provides empirical equations to calculate a temperature–time curve (Tanaka et al. 1997).

The SFPE *Engineering Guide to Fire Exposures to Structural Elements* provides a detailed comparison between alternate methods for compartment fire exposures (SFPE 2004), including the Eurocode design equations (CEN 2001), the method proposed by the Conseil International du Batiment (CIB) (Thomas and Heselden 1972), the parametric equations developed by Magnusson and Thelandersson (1970), and Tanaka et al.'s methods, among others. Hunt et al. (2010) provided a comprehensive evaluation of fifteen different methods for evaluating the compartment fire (gas) temperature history, including the methods specified in the SFPE S.01 standard. Additional discussion on temperature–time curves for enclosure fires can be found in Walton and Thomas (2008) and Thomas (2008). However, the designer must be aware that existing methods to calculate temperature–time curves for enclosure fires are not applicable for all compartment geometries or fuel types/distributions. Thus, the designer should confirm what conditions a given correlation is validated for prior to application of the correlation to calculate an enclosure fire temperature history.

4.3.2 Zone Models

Zone models (e.g., CFAST) solve the equations of mass and energy conservation for one or more control volumes in a compartment. One-zone models are used to define a single temperature–time response for the fire exposure, whereas two-zone models are often used to capture the presence of hot (upper) and cool (lower) gas layers. The two-zone approximation is generally better suited to the conditions in a compartment before flashover occurs.

Zone models typically approximate heat transfer to the boundary surfaces, as well as radiation losses that occur through any vents or openings. Algorithms have been developed to capture more complex phenomena such as mixing between layers and inflow of air through ceiling jets. Quintiere and Wade (2016) provided an overview of the basic conservation equations that are used in zone models, and Walton et al. (2016) briefly discuss computer software that has been developed for compartment fire analysis. Zone models typically require the expertise of a fire protection engineer.

4.3.3 Field Models

Field models [e.g., Fire Dynamics Simulator (FDS)] use computational fluid dynamics (CFD) to study spatially varying fire-induced flows in compartments with complex geometries and fuel conditions. Some field models are based on the Reynolds-averaged Navier–Stokes (RANS) equations, whereas others use large-eddy simulation (LES) (Wang et al. 2009, and McGrattan et al. 2010). Field models solve the conservation equations of mass, momentum, and energy using a mesh of finite difference grids or control volume cells.

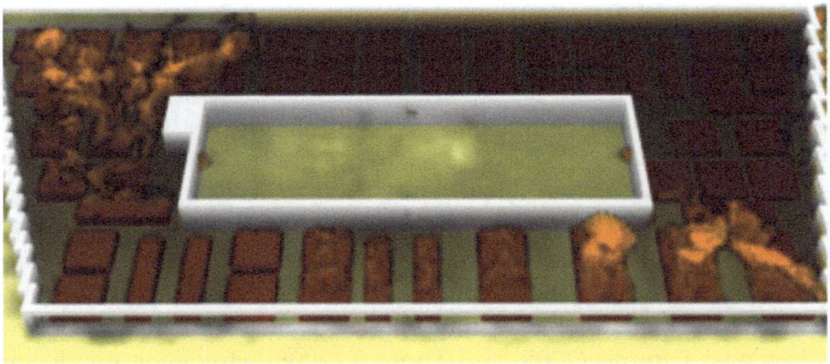

Figure 4-3. Field model simulation of fire exposure.
Source: Courtesy of Simpson Gumpertz & Heger Inc., reprinted with permission.

Field models require considerably more computational power than zone models in exchange for spatial resolution. Figure 4-3 shows an image of a field model that simulates fire spread and smoke development through a large open office floor plan. Because of their complexity, field models require extensive knowledge of fire dynamics, as well as familiarity with the assumptions and limitations embedded in existing software to ensure proper use. Accordingly, field models typically require the expertise of a fire protection engineer

4.3.4 Validation of Fire Models

An extensive validation exercise of the enclosure fire design equations specified in the SFPE S.01 standard can be found in Hunt et al. (2010). As mentioned, the SFPE G.06 guide (SFPE 2011a) provides guidance for selecting an appropriate fire model for a given application. For fire models, it is necessary to demonstrate that the model is appropriate for the intended use and that it is used properly. Determining the suitability of a fire model involves the following steps:

1. *Problem definition*: The design or assessment problem should be clearly defined by identifying the relevant phenomena and key physics, collecting available information, and determining the analysis objectives.
2. *Model selection*: A number of factors should be considered before selecting a particular model for a problem, including computational resources, time limitations, required level of accuracy, and most important, whether or not the governing equations and assumptions in the model are appropriate for the problem of interest.
3. *Verification and validation*: Prior to using a model for a particular problem, the user should determine if the model is capable of generating

a useful result. The formal process by which this is demonstrated is verification and validation (V&V). Verification ensures that the model is working as designed (i.e., the equations are being properly solved). It essentially is a check of the mathematics. Validation is a check of the physics (i.e., the equations are an appropriate description of the fire scenario). Most often, validation takes the form of comparisons with experimental test data.

4. *Uncertainty analysis*: Once the V&V has been conducted, the next step is to focus on the uncertainty that arises in model predictions because of the use of a predictive model. Possible sources of uncertainty include definition of the model space or computational domain, simplifying assumptions (in the application of the model), and the choice of input parameters. The result is a propagation of *error* or uncertainty through the model that should be understood, at minimum, at a qualitative level but preferably quantitatively.

REFERENCES

ASCE/SEI (Structural Engineering Institute). 2016. *Minimum design loads and associated criteria for buildings and other structures.* ASCE/SEI 7. Reston, VA: ASCE.

ASTM International. 2011. *Standard test method for behavior of materials in a vertical tube furnace at 750°C.* ASTM E136. West Conshohocken, PA: ASTM International.

ASTM. 2016. *Standard test method for behavior of materials in a tube furnace with a cone-shaped airflow stabilizer at 750°C.* ASTM E2652. West Conshohocken, PA: ASTM International.

CEN (European Committee for Standardization). 2001. *Eurocode 1—Actions on structures. Part 1–2: General actions—Actions on structures exposed to fire.* ENV 1991-1-2. Brussels: CEN.

CEN. 2010. *Eurocode 3—Design of steel structures. Part 1–2: General rules—Structural fire design.* EN 1993-1-2. Brussels: CEN.

CIB (Conseil International du Batiment). 2001. "Rational fire safety engineering approach to fire resistance of buildings." *CIB Report 269.* Delft, Netherlands: CIB.

Fitzgerald, R., and B. Meacham. 2017. *Fire performance analysis for buildings,* 2nd Ed. Hoboken, NJ: Wiley.

FM Global P0181. 2001. *Fire following earthquake.* Johnston, RI: FM Global.

Hughes Associates. 2006. "Summary of comparisons of test methods." Gaithersburg, MD: Society of Fire Protection Engineers.

Hunt, S., J. Cutonilli, and M. Hurley. 2010. "Evaluation of enclosure temperature empirical models." Gaithersburg, MD: Society of Fire Protection Engineers.

Jonsdottir, A., and G. Rein. 2009. "Out of range." *Fire Risk Manag.* December, 14–17.

Khorasani, N., M. Garlock, and P. Gardoni. 2016. "Probabilistic performance-based evaluation of a tall steel moment resisting frame under post-earthquake fires." *J. Struct. Eng.* 7(3), 193–216.

Kirby, B., D. Wainman, L. Tomlinson, T. Kay, and B. Peacock. 1999. "Natural fires in large-scale compartments." *Int. J. Eng. Perform-based Fire Codes,* 1(2), 43–58.

LaMalva, K. 2017. "Fire design." In *Architect's guidebooks to structures: Special structural topics,* P. McMullin, J. S. Price, and S. Simchuk (eds.). New York: Routledge.

Lattimer, B. 2016. "Heat transfer from fires to surfaces." *SFPE handbook of fire protection engineering,* 5th Ed. Gaithersburg, MD: Society of Fire Protection Engineers.

Law, M., and T. O'Brien. 1983. *Fire safe structural steel.* Washington, DC: American Iron and Steel Institute.

Magnusson, S., and S. Thelandersson. 1970. "Temperature–time curves of complete process of fire development." *Bulletin of division of structural mechanics and concrete construction,* Bull 16. Lund, Sweden: Lund Institute of Technology.

McGrattan, K., H. Baum, R. Rehm, S. Hostikka, and J. Floyd. 2010. "Fire dynamics simulator (version 5) technical reference guide—Volume 1: Mathematical model." *NIST Special Publication 1018-5.* Gaithersburg, MD: NIST.

Mohammadi, J., S. Alyasin, and D. N. Bank. 1992. "Analysis of post-earthquake fire hazard." *Proc., 10th World Conference on Earthquake Engineering,* Balkema, Rotterdam. Madrid: International Association for Earthquake Engineers.

Nelson, H. 1968. "Radiant energy transfer in fire protection engineering problem solving." *Fire Technol.* 4(3), 196–205.

NFPA (National Fire Protection Association). 2012. "Recommended practice for protection of buildings from exterior fire exposures." *NFPA 80A.* Quincy, MA: NFPA.

NFPA. 2016. "Standard for determination of fire loads for use in structural fire protection design." *NFPA 557.* Quincy, MA: NFPA.

NIST. 1995. "Post-earthquake fire and lifelines." *NIST Special Publication 889,* R. M. Chung et al. (eds.). Gaithersburg, MD: NIST.

Quintiere, J., and C. Wade. 2016. "Compartment fire modeling." *SFPE Handbook of Fire Protection Engineering,* 5th Ed. Gaithersburg, MD: Society of Fire Protection Engineers.

Rackauskaite, E., C. Hamel, A. Law, and G. Rein. 2015. "Improved formulation of travelling fires and application to concrete and steel structures." *Structures* 3, 250–260.

Rackauskaite, E., P. Kotsovinos, A. Jeffers, and G. Rein. 2017. "Structural analysis of multi-storey steel frames exposed to travelling fires and traditional design fires." *Eng. Struct.* 150, 271–287.

Scawthorn, C. 2008. "Fire following earthquake." *U.S. Geological Survey Report 2008-1150*. Pasadena, CA: California Geological Survey.

SFPE (Society of Fire Protection Engineers). 1999. *Engineering guide to assessing flame radiation to external targets from pool fires*. Gaithersburg, MD: Society of Fire Protection Engineers.

SFPE. 2004. *Engineering guide to fire exposures to structural elements*. Gaithersburg, MD: Society of Fire Protection Engineers.

SFPE. 2011a. *SFPE G.06: Engineering guidelines for substantiating a fire model for a given application*. Gaithersburg, MD: Society of Fire Protection Engineers.

SFPE. 2011b. *SFPE S.01: Engineering standard on calculating fire exposures to structures*. Gaithersburg, MD: Society of Fire Protection Engineers.

Stern-Gottfried, J., and G. Rein. 2012a. "Travelling fires for structural design—Part I: Literature review." *Fire Saf. J.* 54, 74–85.

Stern-Gottfried, J., and G. Rein. 2012b. "Travelling fires for structural design—Part II: Design methodology." *Fire Saf. J.* 54, 96–112.

Tanaka, T., M. Sato, and T. Wakamatsu. 1997. "Simple formula for ventilation-controlled fire temperatures." *Fire Sci. Technol.* 17(1), 15–27.

Taylor, J. 2003. "Post-earthquake fire in tall buildings and the New Zealand building code." Ph.D. dissertation, School of Engineering, Univ. of Canterbury, Christchurch, New Zealand.

Thomas, I. 2008. "Enclosure fire temperature–time estimation." *SFPE Handbook of Fire Protection Engineering,* 4th Ed. Gaithersburg, MD: Society of Fire Protection Engineers.

Thomas, P., and A. Heselden. 1972. "Fully developed fires in single compartments, A co-operative research program of the Conseil International du Batiment (CIB Report No. 20)." *Fire Research Note No. 923*. Borehamwood, UK: Fire Research Station.

Walton, W., and P. Thomas. 2008. "Estimating temperatures in compartment fires." *SFPE Handbook of Fire Protection Engineering,* 4th Ed. Gaithersburg, MD: Society of Fire Protection Engineers.

Walton, W., P. Thomas, and Y. Ohmiya. 2016. "Estimating temperatures in compartment fires." In *SFPE Handbook of Fire Protection Engineering,* 5th Ed. Gaithersburg, MD: Society of Fire Protection Engineers.

Wang, Y., P. Chatterjee, and S. Dorofeev. 2009. "Large eddy simulation of thermal and fire plumes based on OpenFOAM toolbox." *Proc. Combust. Inst.* 33(2), 2473–2480.

Wang, X., T. C. Hutchinson, G. Hegemier, S. Gunisetty, P. Kamath, and B. Meacham. 2016. "Earthquake and fire performance of a mid-rise cold-formed steel framed building—test program and test results: rapid release (preliminary) report." *SSRP-2016/07*. San Diego, CA: Structural Systems Research Project, Univ. of California.

CHAPTER 5
MATERIAL THERMAL PROPERTIES

For SFED, the designer should determine the relevant thermal properties of the structural and insulating materials. Thermal properties include density, coefficient of thermal expansion, thermal conductivity, emissivity, and specific heat. Most of these thermal properties have significant temperature dependence. However, single values of these properties may be used in certain cases if they would yield conservative thermal response results per the judgment of the designer. The following sections provide guidance for selecting the thermal properties of common structural and insulating materials.

Three primary material thermal properties to consider for thermal analyses are density, specific heat, and thermal conductivity. Density is a measure of the mass of a material per unit volume. Density is relatively easy to measure, and for many materials it is relatively insensitive to temperature. Common units for density are kg/m^3. Specific heat is a measure of the amount of energy required to raise the temperature of a mass of the material. Common units for specific heat are J/(kg·K). Specific heat usually has strong temperature dependence, especially for insulation materials that undergo endothermic reactions under heating (e.g., gypsum board). Thermal conductivity is a measure of a material's ability to conduct heat. Materials with low thermal conductivity may be desirable as insulators. Common units for thermal conductivity are W/(m·K). Thermal conductivity usually has strong temperature dependence, but the dependence is usually not as pronounced as that of the specific heat of certain materials.

In cases where the thermal conductivity decreases or specific heat increases with temperature, ambient-temperature values can be used in most cases. This simplification would result in a greater estimate of temperature rise than using temperature-dependent properties. However, there may be certain cases where such a simplification is not conservative, such as using a higher thermal conductivity for a heat sink (e.g., a column absorbing heat from an adjacent connection). In these cases, the designer should analyze the upper and lower-bound values or use temperature-dependent values.

The physical dimensions of materials change with increased temperature. The coefficient of thermal expansion is a measure of the amount that size changes with temperature. Thermal expansion can induce stresses and/or bending moments in structural elements as they are heated (as specifically described in Section 7.2.1). Common units for thermal expansion are mm/mm/K.

5.1 SPRAY-APPLIED FIRE RESISTANT MATERIALS

Spray-applied fire resistant material (SFRM) products are currently qualified based on standard fire testing. Although these tests provide a fire resistance rating, these ratings have no direct quantitative relationship to the thermal performance of an SFRM product itself. Predictive modeling of thermal response depends critically on having accurate input for the temperature-dependent thermal properties. The properties required for predictive modeling of an SFRM include at a minimum the thermal conductivity, specific heat, emissivity, and density. The enthalpies of reactions and phase changes may also be important if they are not implicitly accounted for in the characterization of the product's specific heat temperature dependence. In that case, enthalpies related to phase change and reaction should be included in the temperature-dependent modified specific heat curve.

SFRM products generally fall into two broad categories: mineral fibers and cementitious materials. However, mixture compositions and the associated thermal properties of SFRM products are usually proprietary to the manufacturer and are not publicly available. It is also possible that the thermal properties of a given SFRM product are not directly characterized by the manufacturer. The magnitude and temperature dependence of SFRM thermal properties can vary significantly among specific products. In addition, standard test methods do not exist for characterizing the thermal properties of SFRM products as a function of temperature. Standard test methods for other materials may be applied, but there may be limitations based on the given test method.

The thermal properties of SFRM products can be highly temperature dependent. NIST conducted testing to characterize the thermal properties of various SFRM products as part of their WTC investigation report (Carino et al. 2005). Temperature-dependent thermal conductivity measurements were performed according to ASTM C1113 (ASTM 2009). Figure 5-1 provides the temperature-dependent thermal conductivity for various SFRM products that were tested by NIST. This test method is based on heating two speci-

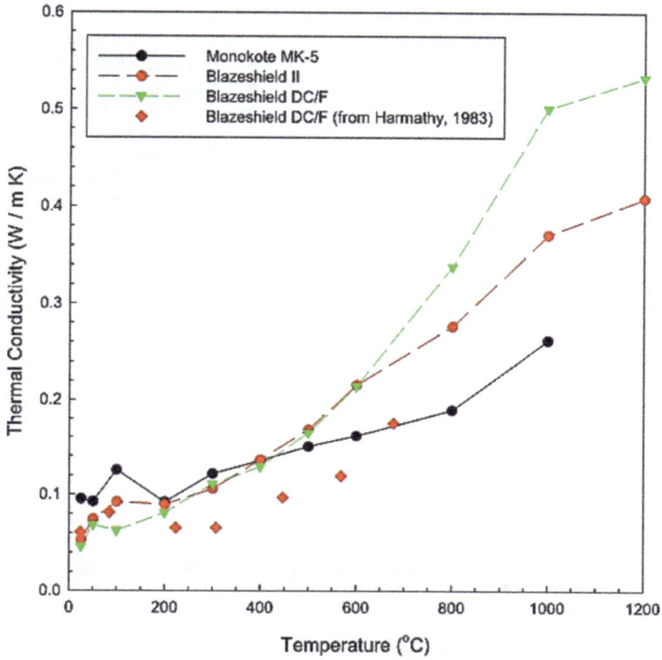

Figure 5-1. Temperature-dependent thermal conductivity of various SFRM products. Source: Carino (2005), reprinted with permission.

mens with a platinum wire placed between them. The variation of the wire's electrical resistance during the test is converted into variation of temperature. Thermal conductivity is then calculated based on the rate of temperature increase of the wire and power input (Carino et al. 2005).

NIST performed differential scanning calorimeter (DSC) testing per ASTM E1269 to measure the temperature-dependent specific heat of various SFRM products (ASTM 2011b). Figure 5-2 provides the temperature-dependent specific heat for various SFRM products that were tested by NIST. Because very small quantities of SFRM products are used in DSC testing, assurance of specimen homogeneity and representativeness of the sample is essential to the validity of the measurements, especially for non-homogeneous materials like SFRMs. In addition, mass loss from the specimen during heating could interfere with the measurements (Carino et al. 2005).

The major disadvantages of using DSC testing to derive the temperature-dependent specific heat of SFRM products include the variable specimen mass during the test exposure due to thermal degradation and the typically small sample size, which may prohibit obtaining a representative volume of the heterogeneous SFRM. However, the former concern has in large

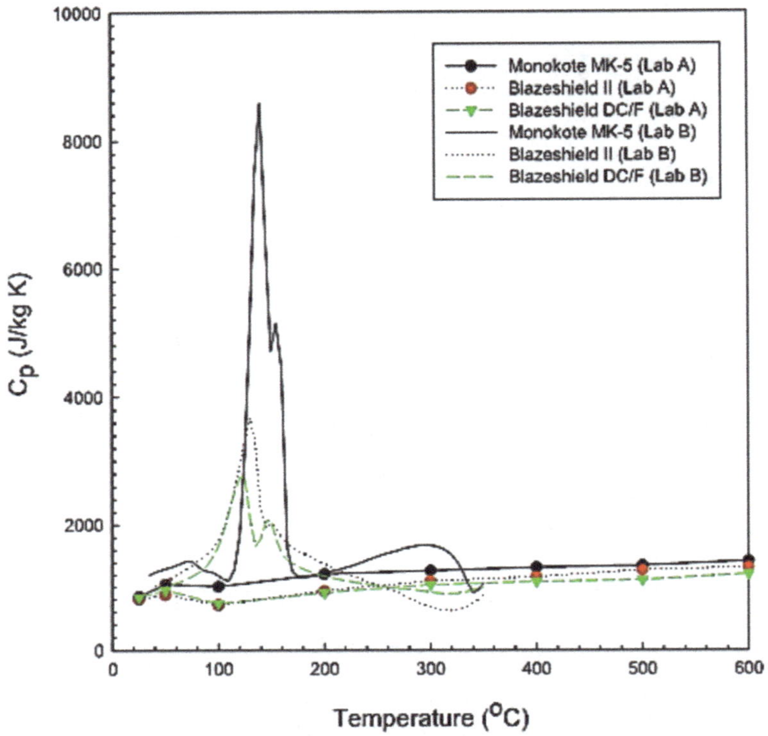

Figure 5-2. Temperature-dependent specific heat of various SFRM products. Source: Carino (2005), reprinted with permission.

part been alleviated in recent years by the availability of commercial simultaneous thermal analysis (STA) units that measure both heat flow and mass during a programmed temperature scan. Some of the newer units also allow for sample sizes on the order of 1 g as opposed to the 50–100 mg sample size limit of conventional DSC equipment, which may alleviate concerns about a representative volume being evaluated (Bentz and Prasad 2007).

NIST performed thermal gravimetric analysis (TGA) and thermal expansion measurements to derive temperature-dependent density measurements for various SFRM products using test methods according to ASTM E1131 (ASTM 2008) and ASTM E228 (ASTM 2011a), respectively. Figure 5-3 provides the temperature-dependent density for various SFRM products that were tested by NIST. Because the SFRM products were not isotropic, separate measurements had to be performed for two orientations. The density values were calculated based on the results of the TGA and thermal

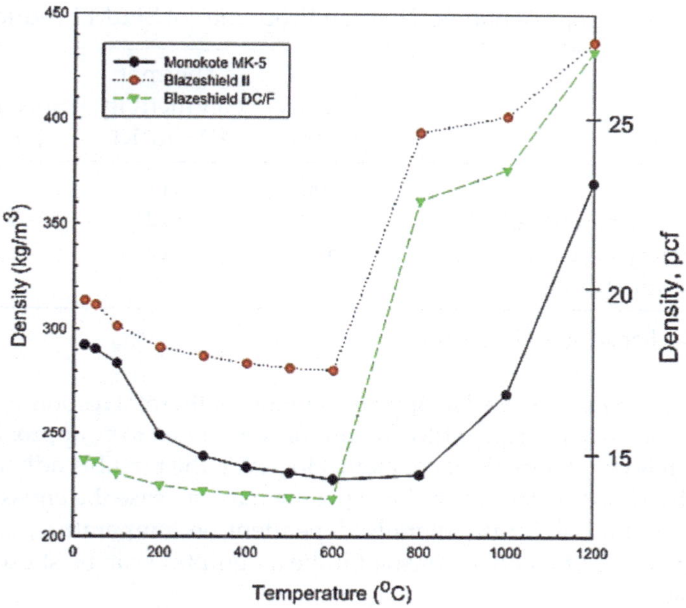

Figure 5-3. Temperature-dependent density of various SFRM products. Source: Carino (2005), reprinted with permission.

expansion testing (Carino et al. 2005). SFRM products typically lose approximately 10–30% of their initial mass during exposure to 1000°C heat owing to dehydration, decarbonation, and other thermal degradation reactions (Bentz et al. 2009).

When conducting TGA testing of SFRMs, it is important to maximize the specimen mass relative to the equipment constraints so that a representative volume of material is analyzed. Analyzing replicate samples can also be used to assess the volumetric heterogeneity in the SFRM. For most conventional TGA equipment, a typical specimen mass will be on the order of 50 mg (Bentz and Prasad 2007).

The thermal conductivity and density of an SFRM depends upon how it is sprayed or applied, and the results will vary between installations. In general, the thermal conductivity of a porous material is a complex function of bulk density, porosity, and other material properties. Recent attempts to use existing predictive methods to estimate thermal conductivities of porous media for SFRM products show promise, but they may have certain inaccuracies (Carino et al. 2005).

Table 5-1 provides representative thermal properties of SFRM products (Buchanan and Abu 2017). These representative values do not consider the temperature dependence or the specific formulation of a given SFRM product,

Table 5-1. Representative Thermal Properties of SFRM Products.

SFRM Type	Density [kg/m^3]	Thermal conductivity [W/m*K]	Specific heat [J/kg*K]
Mineral fiber	300	0.12	1200
Perlite or vermiculite plaster	350	0.12	1200
High-density perlite or vermiculite plaster	550	0.12	1200

Source: Buchanan and Abu (2017).

so they should only be used as approximations for thermal response calculations. The emissivity of an SFRM is commonly assumed to be approximately 0.9, which is characteristic of concrete. However, the possible adherence of soot to the SFRM surface from fire exposure may increase the emissivity of the surface. Emissivity is generally dependent on temperature, although emissivity can be treated as constant if the assumption can be shown to be conservative.

For the specific SFRM products studied, it is reasonable for the designer to use the temperature-dependent thermal properties reported by Carino et al. (2005). Otherwise, if values are not available in the literature, manufacturers should be consulted by designers conducting SFED if temperature-dependent thermal properties of the SFRM products are sought. Alternatively, the designer may obtain values through calibration to furnace testing results if deemed conservative. This would involve comparing thermal modeling results to temperature histories obtained through furnace testing. This approach should be conducted with caution because each of the individual temperature-dependent thermal properties would be coupled.

5.2 GYPSUM BOARD PRODUCTS

Similar to SFRM, gypsum board products are currently qualified based on standard fire testing. Gypsum board is produced by mixing water with plaster of Paris (calcium sulfate hemihydrate) or with Keene's cement (calcium sulfate anhydrite). Gypsum board products are used extensively in the building construction industry in the form of wallboard, form board, and sheathing. Gypsum board, based on its composition and performance, is classified into various types, such as regular, Type X, and improved Type X (i.e., Type C). A gypsum board with naturally occurring fire resistance from the gypsum core is defined as regular gypsum. When the core of the gypsum board is modified with special core additives or with enhanced additional properties to improve the natural fire resistance from regular gypsum board, it is classified as Type X or Type C gypsum board (Kodur and Harmathy 2016). Lastly, gypsum board may be classified as moisture resistant with certain applications such as exterior facades.

There may be significant variation in the thermal properties of gypsum board based on the type and the formulation of the core, which varies from one manufacturer to another. The variation with temperature of the specific heat of pure gypsum was derived by Harmathy (1961) based on information from West and Sutton (1954) and Ljunggren (1960). The thermal conductivity of gypsum products is difficult to assess because of the large variations in their porosities. A typical value for plaster boards of about 700 kg/m^3 is 0.25 W/m·K.

NIST conducted thermophysical testing of four thicknesses of gypsum board ranging from ½ in. to 1 in. (Carino et al. 2005). The thermal conductivity of the gypsum board samples was measured using the heated probe technique per ASTM D5334 (ASTM 2014). Figure 5-4 shows the temperature-dependent thermal conductivity of the gypsum board samples tested by NIST. NIST used DSC testing per ASTM E1269 (ASTM 2011b) to derive the specific heat of the gypsum board samples (Figure 5-5). It was observed that as the gypsum board samples were heated, the hydrated water within them was liberated in two endothermic reactions. NIST used an STA to measure mass changes and a dual push-rod dilatometer per ASTM E228 (ASTM 2011a) in order to derive the temperature-dependent density of the gypsum board samples (Figure 5-6).

In addition to testing by NIST, Kodur and Harmathy (2016) provided the temperature dependence of the thermal conductivity and specific heat of Type X gypsum board derived from testing performed by Sultan (1996). Lastly, Table 5-2 provides representative thermal properties of board products (Buchanan and Abu 2017). These representative values do not consider the temperature dependence or specific formulation of a given board product, so they should only be used as approximations for thermal response calculations.

For the specific gypsum board products studied, it is reasonable for the designer to use the temperature-dependent thermal properties reported by Carino et al. (2005) and Kodur and Harmathy (2016). Otherwise, if values are not available in the literature, manufacturers should be consulted by designers conducting SFED if the temperature-dependent thermal properties of gypsum board products are sought. Alternatively, the designer may obtain values through calibration to furnace testing results if deemed conservative. This would involve comparing thermal modeling results to temperature histories obtained through furnace testing. This approach should be conducted with caution because each of the individual temperature-dependent thermal properties would be coupled.

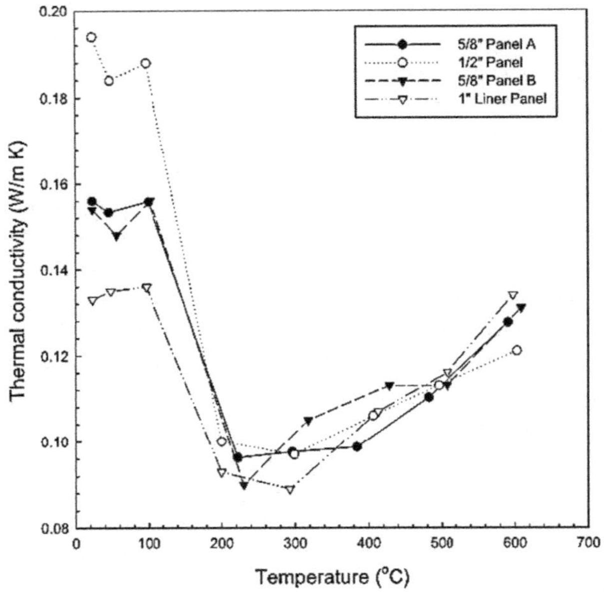

Figure 5-4. Temperature-dependent thermal conductivity of gypsum board samples. Source: Carino (2005), reprinted with permission.

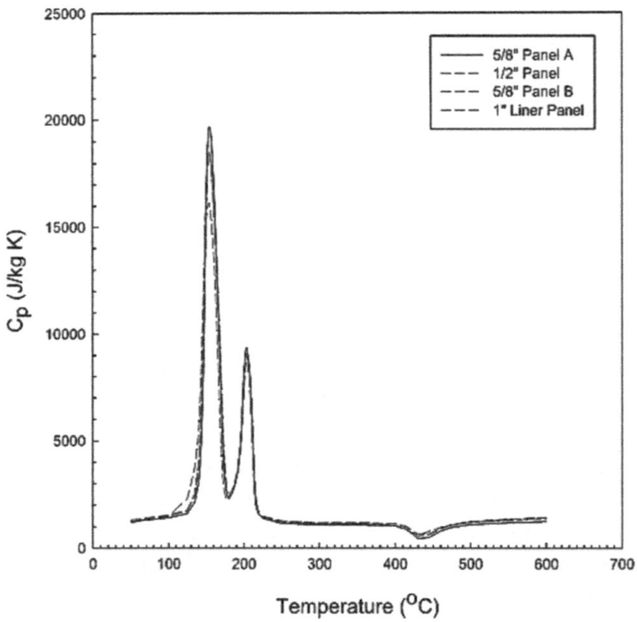

Figure 5-5. Temperature-dependent specific heat of gypsum board samples. Source: Carino (2005), reprinted with permission.

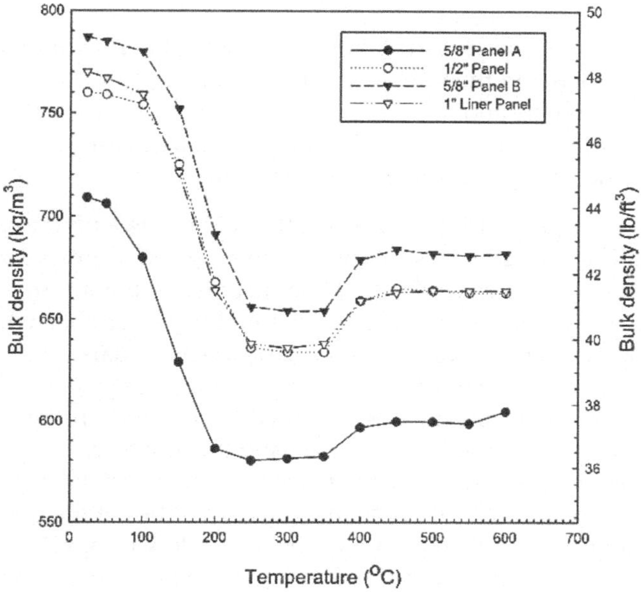

Figure 5-6. Temperature-dependent density of gypsum board samples. Source: Carino (2005), reprinted with permission.

Table 5-2. Representative Thermal Properties of Board Products.

Board type	Density [kg/m^3]	Thermal conductivity [W/m·K]	Specific heat [J/kg·K]
Fiber silicate or fiber calcium silicate	600	0.15	1200
Gypsum plaster	800	0.2	1700

Source: Buchanan and Abu (2017).

5.3 INTUMESCENT COATINGS

The physical and chemical behavior of intumescent coatings in the presence of fire is highly complex and most often proprietary. The base material of an intumescent coating (epoxy matrix or thin-film material for intumescent paints) has a relatively high thermal conductivity, whereas the reaction product (the char layer formed by swelling) acts as a thermal barrier with a low thermal conductivity. The density is reduced during the process of swelling (Opstad 2010). The resulting char layer thickness can be between 5 and 100 times that of the base material (Butler 1997). The chemical compounds of intumescent systems are classified into four categories:

- A carbonization agent,
- A carbon-rich polyhydric compound that influences the amount of char formed and the rate of char formation,
- An acid source, and
- A foaming agent, which during degradation releases nonflammable gases such as carbon dioxide and ammonia.

At higher temperatures (200–300°C), the acid reacts with the carbonization agent. The formed gases expand, beginning the intumescence in the form of carbonaceous char. The low conductivity of the intumescent char results from the pockets of trapped gas within the porous char, which act as a blowing agent to the solid material (Mesquita et al. 2009).

An intumescent coating protects the underlying substrate from heating through two mechanisms: heat is absorbed by the endothermic chemical reactions that produce the formed gases, and the low thermal conductivity of the porous char provides an insulating layer (Butler et al. 1995). The timing of chemical events is critical to the protective qualities of the final char. Of particular importance is the decomposition of the blowing agent with respect to the melt viscosity. If gasification takes place when the melt viscosity is too high, the formed gas growth will be strongly opposed, and the gas will tend to diffuse through the mixture without generating bubbles. If the melt viscosity is too low, large bubbles will form, resulting in a fragile and ineffective char (Butler 1997).

Because of the complexity of the reactions, the design of intumescent materials has traditionally proceeded by means of systematic testing of a variety of chemical formulations (Butler et al. 1995). The swelling process and its mechanism are difficult to represent mathematically. Moreover, tabulated thermal data from standard fire tests are not directly applicable to realistic fires considered in SFED. Most heat transfer simulations rely on a constant geometry mesh. The conductivity versus temperature curve must therefore take into account the phase change from a solid material to a porous char and the increased thickness. Further, this process is dependent on thickness, mass of substrate, and the intensity and duration of the heat flux (Opstad 2010).

The designer should understand the behavior, as well as the physical and mechanical reactions, of the intumescent material being analyzed. Clearly, intumescent materials present considerable difficulties to designers seeking to analytically model its behavior as part of SFED. Swelling is central to the fire protective capabilities of intumescent coatings, and a fundamental understanding of the mechanisms that cause expansion is important. In particular, the effect of growing bubbles on the temperature field cannot be neglected (Butler 1997). It has been proposed that instead of putting effort into modeling the swelling process, designers could assume initial conditions with respect to density and specific heat and include all the changes to the property through the thermal conductivity versus temperature curve (Opstad 2010). If this method is used, conservative assumptions should be enacted per the judgment of the designer given the complexity and in-situ dependence of the behavior (e.g., substrate thickness). Wang et al. (2013) also discuss the thermal properties of intumescent coatings.

Designers should use caution when deriving temperature-dependent properties of intumescent coating products from the literature. In all cases, it is recommended that the designer contact the product manufacturer for the proper characterization of thermal properties for an SFED.

5.4 STEEL

The thermal properties of steel are well established and fairly consistent across most common grades of structural steel. The Eurocode 3 equations are most commonly cited in the literature (CEN 2010), although different equations have been proposed by the ASCE manual *Structural Fire Protection* (ASCE 1992). A comparison between the two models shows that there are slight differences between the two models (Kodur et al. 2010). Both methods are presented in this section, and the slight differences should be understood by the designer.

The density of steel can be taken as 490 lb/ft^3 (7,850 kg/m^3), and it remains constant with temperature. The thermal conductivity and specific heat of steel are dependent on temperature and can be modeled using the Eurocode equations (CEN 2010). In particular, the conductivity k_s (in W/m·K) of steel at temperature T (in Celsius) is given as

$$k_s = \begin{cases} 54 - 3.33\times10^{-2}T & \text{for } 20°\text{C} \le T \le 800°\text{C} \\ 27.3 & \text{for } 800°\text{C} < T \le 1200°\text{C} \end{cases} \tag{5-1}$$

The variation of thermal conductivity with temperature is shown in Figure 5-7. For most applications, the consideration of the upper bound thermal conductivity would yield conservative results.

The specific heat c_s (in J/kg·K) of steel at temperature T (in Celsius) is given as (CEN 2010)

$$c_s = \begin{cases} 425 + 7.73\times10^{-1}T - 1.69\times10^{-3}T^2 + 2.22\times10^{-6}T^3 & \text{for } 20°\text{C} \le T \le 600°\text{C} \\ 666 + 13002/(738 - T) & \text{for } 600°\text{C} < T \le 735°\text{C} \\ 545 + 17820/(T - 731) & \text{for } 735°\text{C} < T \le 900°\text{C} \\ 650 & \text{for } 900°\text{C} < T \le 1200°\text{C} \end{cases} \tag{5-2}$$

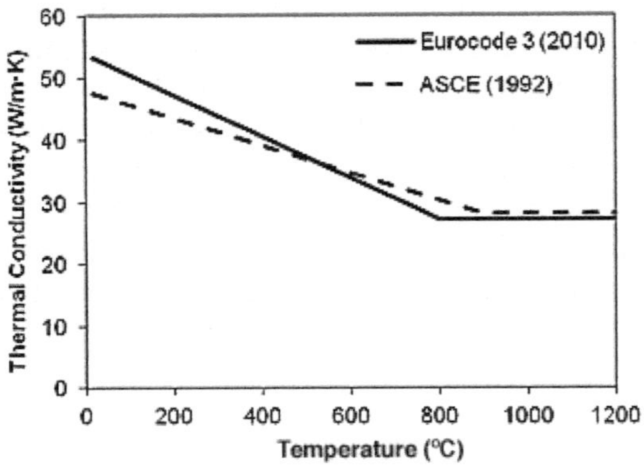

Figure 5-7. Thermal conductivity of steel with temperature.

The thermal expansion of steel is generally expressed in terms of the thermal strain (i.e., the change in length Δl due to the temperature increase divided by the original element length l). The coefficient of thermal expansion is dependent on temperature, and the thermal strain is typically nonlinear as a function of temperature. For structural steel, thermal expansion can be modeled using the Eurocode equations (CEN 2010). In particular, the thermal strain is given as

$$\Delta l / l = \begin{cases} 1.2 \times 10^{-5} T + 0.4 \times 10^{-8} T^2 - 2.416 \times 10^{-4} & \text{for } 20°\text{C} \le T \le 750°\text{C} \\ 1.1 \times 10^{-2} & \text{for } 750°\text{C} < T \le 860°\text{C} \\ 2 \times 10^{-5} T - 6.2 \times 10^{-3} & \text{for } 860°\text{C} < T \le 1200°\text{C} \end{cases} \tag{5-3}$$

Figures 5-8 and 5-9 illustrate the variation of the specific heat and thermal strain of steel with temperature, respectively (CEN 2010 and ASCE 1992). For most applications, the consideration of the lower-bound specific heat would yield conservative results.

For protected steel members, the emissivity should be specified for the protective insulation rather than for the steel because of shielding that is provided by the insulation. For unprotected steel members, the emissivity depends on the surface coating (e.g., oxidized, galvanized, polished, or painted). Consequently, the emissivity of unprotected steel at room temperature ranges from 0.20 to 0.90 (FABIG 1993). Under fire exposure, soot may adhere to the surface of steel. Hence, it would be conservative to consider an emissivity value toward the higher end of the provided range. The emissivity is generally dependent on temperature, although emissivity can be treated as constant if the assumption can be shown to be conservative.

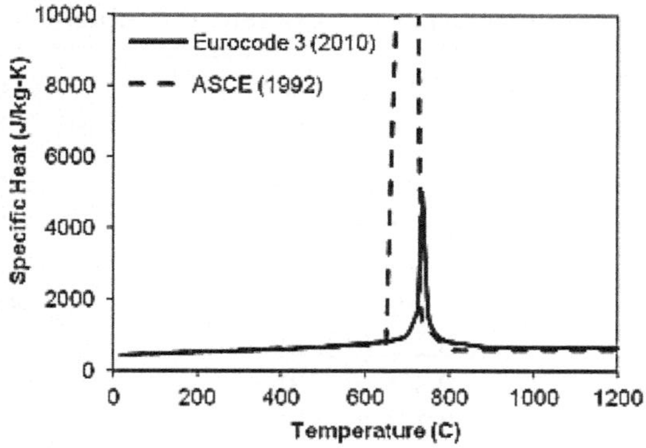

Figure 5-8. Specific heat of steel with temperature.

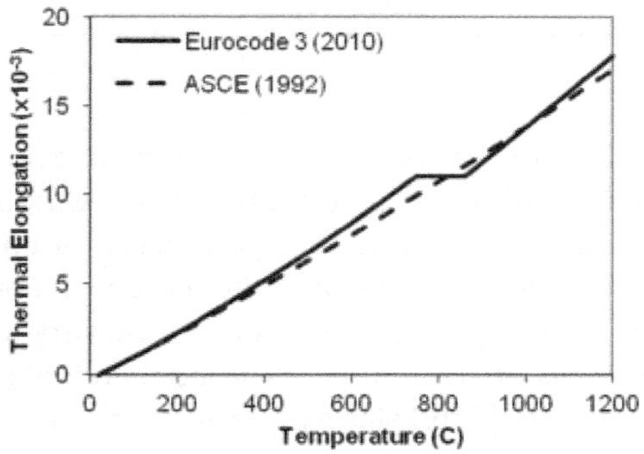

Figure 5-9. Thermal elongation of steel with temperature.

5.5 CONCRETE

The thermal properties of concrete depend on a number of factors, including the porosity, the moisture content, and the constituent materials (Szoke 2006). Two principal types of aggregate used in normal-weight concrete are silicate (siliceous) and carbonate aggregate. Material data for concrete are typically reported for small-scale specimens, whereas the thermal performance of a full-scale structure may deviate from small-scale specimens because of the effects of moisture migration (Milke et al. 2010). Also, the thermal properties of concrete vary widely based on the density and type of aggregate (Harmathy 1993), but no discernible difference exists between normal-strength and high-strength concretes (Burg and Ost 1994, Van Geem et al. 1996). Hence, designers should carefully consider these factors when choosing conservative values for SFED.

Table 5-3. Thermal Properties of Concrete at Ambient Temperature.

Density ρ_c, lb/ft^3 (kg/m^3)	Thermal conductivity k_c, Btu/(h·ft·F) (W/m·K)	Specific heat c_c, Btu/(lb·F) (kJ/kg·K)
50 (800)	0.113 (0.196)	0.21 (0.88)
60 (960)	0.138 (0.239)	0.21 (0.88)
70 (1,120)	0.169 (0.292)	0.21 (0.88)
80 (1,280)	0.206 (0.357)	0.21 (0.88)
90 (1,440)	0.252 (0.436)	0.21 (0.88)
100 (1,600)	0.308 (0.533)	0.21 (0.88)
110 (1,760)	0.376 (0.651)	0.21 (0.88)
120 (1,920)	0.459 (0.794)	0.21 (0.88)
130 (2,080)	0.563 (0.974)	0.22 (0.92)
140 (2,240)	0.685 (1.19)	0.22 (0.92)
150 (2,400)	0.836 (1.45)	0.22 (0.92)

Source: ACI/TMS (2007).

At a minimum, the density, conductivity, emissivity, and specific heat for concrete are needed to predict its thermal response. Although there would be some level of moisture evaporation, the density of concrete does not change very much with temperature, and therefore it can be treated as constant. The conductivity and specific heat of concrete do vary with temperature, although these values are often assumed to be independent of temperature in design calculations. The ACI 216.1 standard provides the thermal conductivity and specific heat at ambient temperature for concrete with densities ranging from 50 lb/ft^3 (800 kg/m^3) to 150 lb/ft^3 (2,400 kg/m^3) for analyses in which material data is not readily available (ACI/TMS 2007), as shown in Table 5-3.

To account for the temperature-dependent thermal properties of concrete, the Eurocode 2 equations for density, conductivity, and specific heat may be used (CEN 2005). To account for the reduction in density due to free water loss (i.e., evaporation of water not bound by hydration), the density ρ_c (in kg/m^3) at temperature T (in Celsius) can be calculated by Equation 5-4.

$$\rho_c = \begin{cases} \rho_{c,20} & \text{for } 20°\text{C} \le T \le 115°\text{C} \\ \rho_{c,20}\left[1 - 0.02(T - 115)/85\right] & \text{for } 115°\text{C} < T \le 200°\text{C} \\ \rho_{c,20}\left[0.98 - 0.03(T - 200)/200\right] & \text{for } 200°\text{C} < T \le 400°\text{C} \\ \rho_{c,20}\left[0.95 - 0.07(T - 400)/800\right] & \text{for } 400°\text{C} < T \le 1200°\text{C} \end{cases} \quad (5\text{-}4)$$

The parameter $\rho_{c,20}$ is the density of concrete at ambient temperature (in kg/m^3). The density of normal-weight concrete with temperature is shown in Figure 5-10.

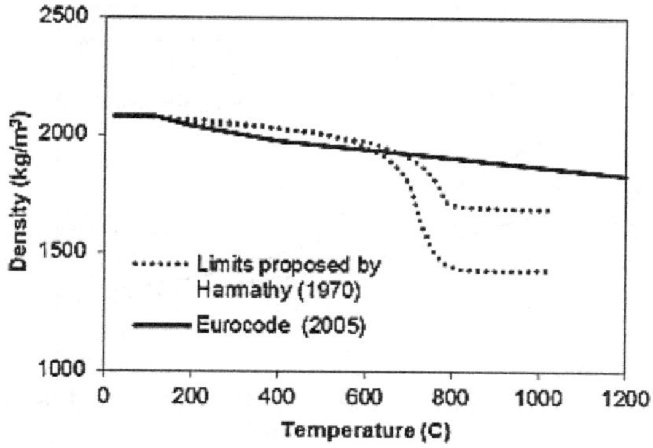

Figure 5-10. Density of normal-weight concrete with temperature.

The thermal conductivity k_c (in W/m·K) of normal-weight concrete at temperature T (in Celsius) can be determined between the upper and lower limits given in Equation 5-5.

$$k_c = \begin{cases} 2 - 0.2451(T/100) + 0.0107(T/100)^2 & \text{upper limit} \\ 1.36 - 0.136(T/100) + 0.0057(T/100)^2 & \text{lower limit} \end{cases} \tag{5-5}$$

For most applications, the consideration of the upper-limit thermal conductivity would yield conservative results. For lightweight concrete with densities between 100 lb/ft³ (1,600 kg/m³) and 125 lb/ft³ (2,000 kg/m³), the conductivity k_c (in W/m·K) is given as

$$k_c = \begin{cases} 1 - (T/1600) & \text{for } 20^\circ\text{C} \le T \le 800^\circ\text{C} \\ 0.5 & \text{for } 800^\circ\text{C} < T \le 1200^\circ\text{C} \end{cases} \tag{5-6}$$

The variation of conductivity with temperature is shown in Figure 5-11 for normal-weight and lightweight concretes.

For normal-weight concrete, the specific heat of dry concrete c_c (in J/kg·K) at temperature T (in Celsius) is given as

$$c_c = \begin{cases} 900 & \text{for } 20^\circ\text{C} \le T \le 100^\circ\text{C} \\ 900 + (T - 100) & \text{for } 100^\circ\text{C} < T \le 200^\circ\text{C} \\ 1000 + \dfrac{T - 200}{2} & \text{for } 200^\circ\text{C} < T \le 400^\circ\text{C} \\ 1100 & \text{for } 400^\circ\text{C} < T \le 1200^\circ\text{C} \end{cases} \tag{5-7}$$

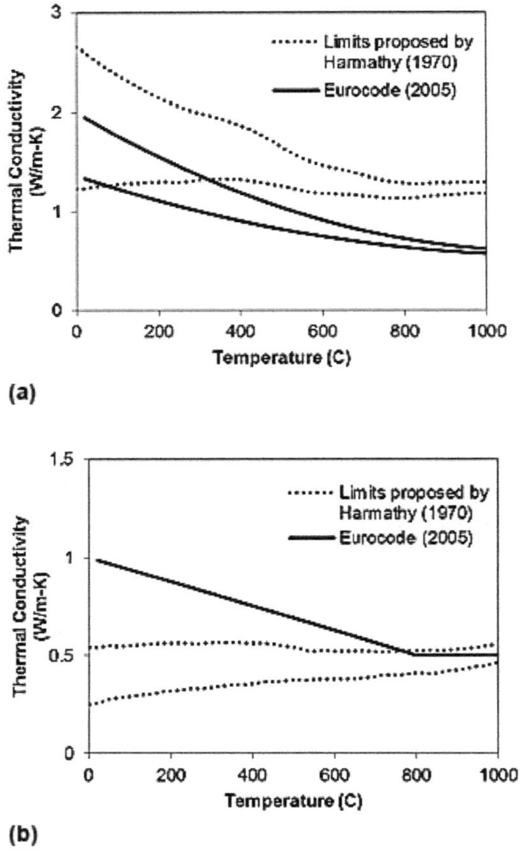

Figure 5-11. Thermal conductivity of concrete with temperature: (a) normal weight concrete, (b) lightweight concrete.

When the moisture content is not modeled explicitly, the specific heat of concrete c_c^* (in J/kg·K) can be considered as follows for different moisture content values (u) at 115°C:

$$c_c^* = \begin{cases} 1470 & \text{for } u = 1.5\% \\ 2020 & \text{for } u = 3.0\% \\ 5600 & \text{for } u = 10.0\% \end{cases} \tag{5-8}$$

The parameter u is the moisture content. For lightweight concrete with densities between 100 lb/ft^3 (1,600 kg/m^3) and 125 lb/ft^3 (2,000 kg/m^3), the specific heat is assumed to be independent of temperature and of magnitude 840 J/kg·K.

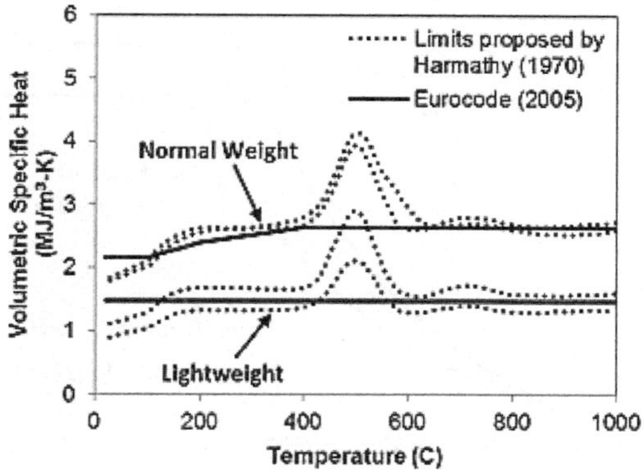

Figure 5-12. Volumetric specific heat of concrete with temperature.

The volumetric specific heat for dry concrete ($u = 0\%$) is plotted in Figure 5-12 for normal-weight and lightweight concrete. For most applications, the consideration of specific heat for dry concrete would yield conservative results. Also, use of the lower-bound curves from Eurocode 2 (CEN 2005) would yield conservative results.

The thermal expansion of concrete is dependent on temperature and aggregate type. For normal-weight concrete with siliceous aggregate, Eurocode 2 (CEN 2005) gives the thermal strain as

$$\Delta l / l = \begin{cases} -1.8 \times 10^{-4} + 9 \times 10^{-6} T + 2.3 \times 10^{-11} T^3 & \text{for } 20°\text{C} \le T \le 700°\text{C} \\ 1.4 \times 10^{-2} & \text{for } 700°\text{C} < T \le 1200°\text{C} \end{cases} \tag{5-9}$$

For normal-weight concrete with calcareous aggregates, the thermal strain is given as

$$\Delta l / l = \begin{cases} -1.2 \times 10^{-4} + 6 \times 10^{-6} T + 1.4 \times 10^{-11} T^3 & \text{for } 20°\text{C} \le T \le 805°\text{C} \\ 1.2 \times 10^{-2} & \text{for } 805°\text{C} < T \le 1200°\text{C} \end{cases} \tag{5-10}$$

For lightweight concrete, the thermal strain is given as

$$\Delta l / l = 8 \times 10^{-6} (T - 20) \tag{5-11}$$

The thermal expansion of concrete is plotted against temperature in Figure 5-13 for normal-weight and lightweight concrete.

The emissivity of concrete ranges between 0.85 and 0.95 and is consistent for both normal and lightweight concretes (FABIG 1993). Under fire exposure, soot may adhere to the surface of concrete. Hence, it would be

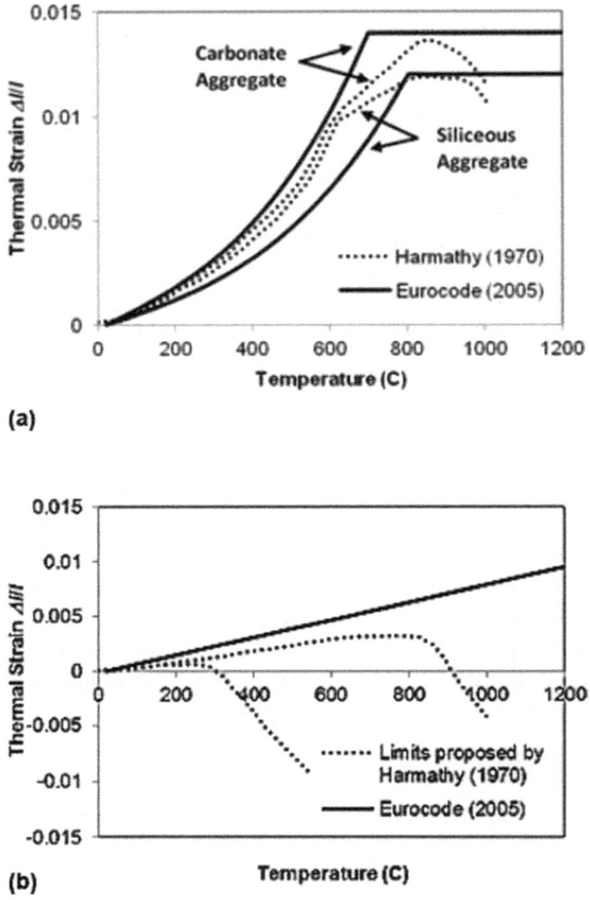

Figure 5-13. Thermal expansion of concrete with temperature, (a) normal-weight concrete, (b) lightweight concrete.

conservative to consider an emissivity value toward the higher end of the provided range. The emissivity is generally dependent on temperature, although emissivity can be treated as constant if the assumption can be shown to be conservative.

5.6 MASONRY

Structural masonry is an assembly of masonry units bonded together with mortar. Common types of masonry units include clay, calcium silicate, autoclaved aerated concrete (AAC), and lightweight concrete aggregate. The most significant material property characterizing the thermal performance of masonry is density rather than the type of brick, as the density is a measure of the porosity of the brick, which governs the thermal properties of masonry at elevated temperatures (Purkiss 1996).

In the absence of specific data on the thermal properties of masonry, the ASCE/SEI 29-05 standard provides thermal conductivity values for concrete and clay masonry, as shown in Tables 5-4 and 5-5, respectively (ASCE/SEI 2005). Similarly, ACI/TMS 216 provides design values for thermal conductivity of 2.16 W/(m·K) and 3.89 W/(m·K) for clay masonry with densities of 120 pcf and 130 pcf, respectively (ACI/TMS 2007). ASCE/SEI 29-05 provides design values for specific heat of concrete and masonry units as 1.05 kJ/(kg·K) and 1.00 kJ/(kg·K), respectively. Eurocode 6 also provides thermal properties for masonry (CEN 2003). Figures 5-14 through 5-17 plot the temperature-dependent thermal conductivity, specific heat, and density for AAC units, calcium silicate units, light aggregate concrete units, and clay units, respectively.

The emissivity of masonry is commonly assumed to be approximately 0.9, which is characteristic of concrete and clay.

Table 5-4. Thermal Conductivity of Concrete Masonry at Ambient Temperature.

Masonry density [lb/ft^3]	Thermal conductivity [W/(m·K]
80	0.36
85	0.40
90	0.43
95	0.48
100	0.54
105	0.59
110	0.66
115	0.73
120	0.80
125	0.88
130	0.97
135	1.07
140	1.19
145	1.32
150	1.45

Source: ASCE/SEI (2005).

Table 5-5. Thermal Conductivity of Clay Masonry at Ambient Temperature.

Masonry density [lb/ft³]	Thermal conductivity [W/(m·K)]
120	2.16
130	3.89

Source: ASCE/SEI (2005).

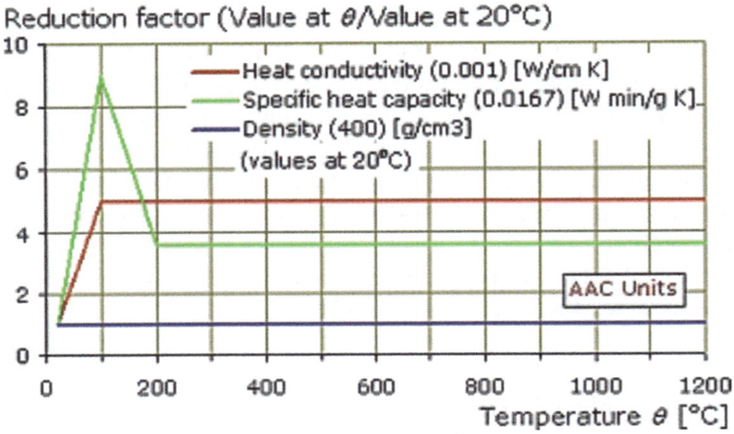

Figure 5-14. Thermal properties of AAC units with density range of 400–600 kg/m³.
Source: Bailey (2016).

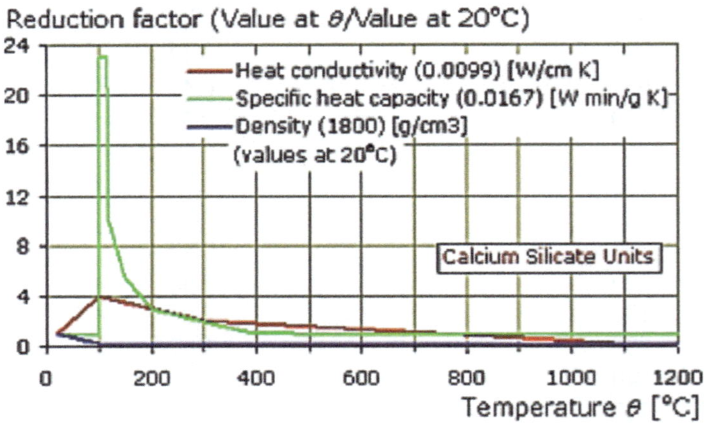

Figure 5-15. Thermal properties of calcium silicate units with density range of 1,600–2,000 kg/m³.
Source: Bailey (2016).

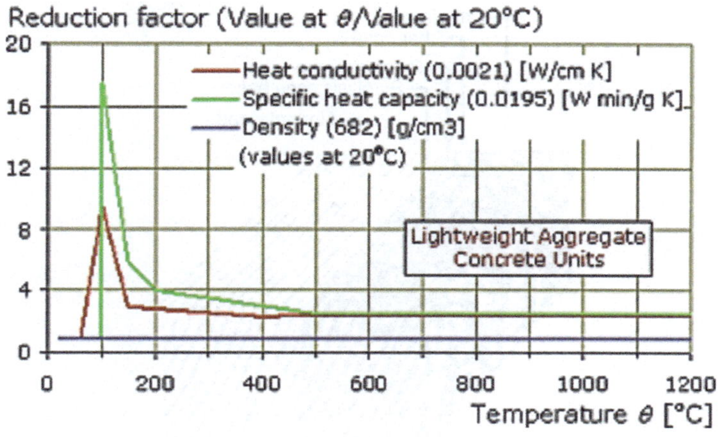

Figure 5-16. Thermal properties of lightweight aggregate concrete units with density range of 600–1,000 kg/m^3.
Source: Bailey (2016).

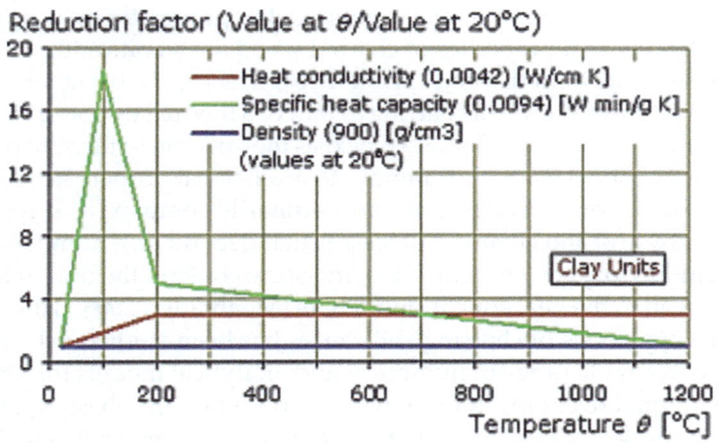

Figure 5-17. Thermal properties of clay units with density range of 900–1,200 kg/m^3.
Source: Bailey (2016).

5.7 TIMBER

The thermal analysis of timber structures requires special consideration of the thermal decomposition of wood under heating. As shown in Figure 5-18, thermal decomposition begins with a moving pyrolysis front that transforms the wood to char through a process that results in mass loss, discoloration, and the emission of gaseous byproducts (Buchanan and Abu 2017).

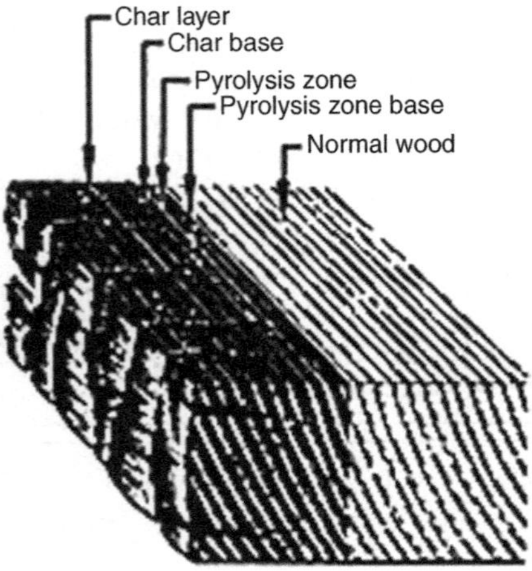

Figure 5-18. Char layer and pyrolysis zone in wood.
Source: Schaffer (1967), reprinted with permission.

Knowledge of the char extent is required for proper thermal analyses because the thermal properties of charred wood are significantly different from those of uncharred wood (White 2016). Hence, the designer should analyze the expected level of charring as part of a thermal response analysis. The char rates of wood are discussed next as they pertain to thermal properties and in Section 8.5.4 as they pertain to mechanical properties.

The progression of pyrolysis is extraordinarily complex, as it typically involves physical and chemical processes that depend on a number of factors, including oxygen concentration, moisture content, the orientation of the specimen to the heat source, and others (Moghtaderi 2006, Babrauskas 2001). The pyrolysis of charring substances has been studied extensively, resulting in a wide range of numerical and analytical models for predicting the thermal decomposition of wood exposed to fire. Existing models range in complexity from simple calculation models (Benichou and Sultan 2000) to integral models that approximate the pyrolysis behavior using partial differential equations (Moghtaderi et al. 1997, Spearpoint and Quintiere 2000) to comprehensive models that account for both diffusion and chemical kinetics (Di Blasi 1993, Antal and Varhegyi 1995, Moghtaderi 2006).

Char develops between 536°F (280°C) and 608°F (320°C), with a temperature of 550°F (288°C) typically used in locating the pyrolysis front using embedded thermocouples (Schaffer 1984). Under standard fire exposure, the pyrolysis front moves at a fairly constant rate after approximately 45 min. Although the rate of charring is effectively *nonlinear* because of the insulating effect of the char layer, the charring rate at one hour under a standard fire exposure is often used as the *nominal* char rate and is used as the reference char rate in nonlinear charring models, such as those used in the *National Design Specification for Wood Construction* (AWC 2012). Moisture content is a major factor affecting char rate. Other factors such as wood species, density, orientation of grain, and inherent variability of wood as a natural material have a lesser influence.

In general, the nominal char rate, β_n, for softwoods ranges from 0.024 to 0.033 in./min (0.61 to 0.84 mm/min), whereas the nominal charring rate for hardwoods is less than 0.021 in./min (0.53 mm/min). Note that these values apply for solid wood members, whereas the nominal charring rate for glue-laminated (*glulam*) timber and laminated veneer lumber (LVL) may differ. The charring rate normal to the wood grain is roughly one-half of the charring rate parallel to the wood grain (Schaffer 1984). Charring rate data for selected wood species subjected to standard fire testing and to constant heat flux can be found in the *Wood Handbook* (FPL 2010). Also, White (2016) gives a review of charring rate models.

NDS has recently incorporated a nonlinear char rate model (AWC 2012). This model was based on work by White (1988). The depth (in.) of the layer of char under the standard fire exposure can be determined by

$$d_{char} = 1.2\beta_n \cdot t^{0.813} \tag{5-12}$$

The parameter β_n is the nominal char rate (in./h) and t is the time (h). Although this is a one-dimensional relationship, the 1.2 factor in the equation above is used to account for two-dimensional charring at corners and strength loss in the elevated-temperature region adjacent to the char front. For example, the char layer of a timber member of width b and depth h exposed to heating on three sides (e.g., top of beam in contact with floor above) can be approximated by an equivalent char depth, as illustrated in Figure 5-19. For nonstandard fire exposure, specific equations may be used, as discussed by White (2016).

In terms of uncharred wood, the density of various wood species can be found in the *Wood Handbook* (FPL 2010). Most species have an oven-dried density in the range of 20 to 45 lb/ft^3 (320 to 720 kg/m^3) (White 2016). The moisture content of wood greatly affects its thermal properties. However, the moisture content is dependent on temperature and the relative humidity of the environment. Under the conditions of standard fire testing (i.e., at 50% relative humidity), the equilibrium moisture content is approximately 9% (White 2016).

As temperature increases, the internal moisture evaporates, which results in a decrease in density. Eurocode 5 accounts for the reduction in density using the ratio provided in Table 5-6, where M is the moisture content (CEN 2004). The reduction factor is plotted in Figure 5-20 for a moisture content of 9% (i.e., $M = 0.09$) by weight. The designer may interpolate between values provided in Table 5-6.

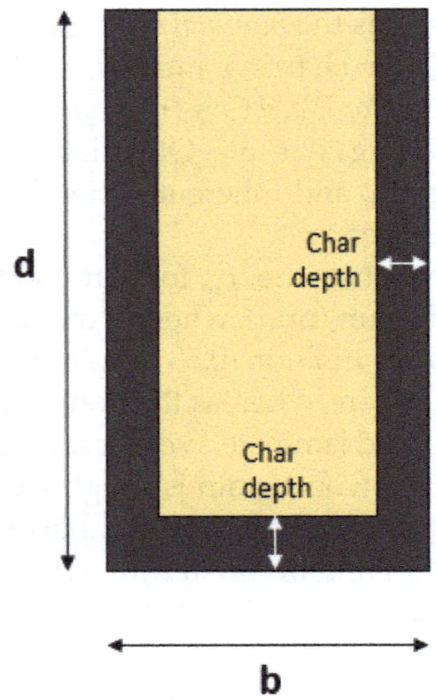

Figure 5-19. Approximation of the char layer in a timber member exposed to heating on three sides.

Table 5-6. Thermal Properties of Wood and Char.

Temperature (°C)	Density ratio (-)	Conductivity (W/m·K)	Specific heat (kJ/kg·K)
20	$1+M$	0.12	1.53
99	$1+M$	—	1.77
99	—	—	13.6
120	—	—	13.5
120	1	—	2.12
200	1	0.15	2
250	0.93	—	1.62
300	0.76	—	0.71
350	0.52	0.07	0.85
400	0.38	—	1
500	—	0.09	—
600	0.28	—	1.4
800	0.26	0.35	1.65
1200	0	1.5	1.65

Source: CEN (2004).

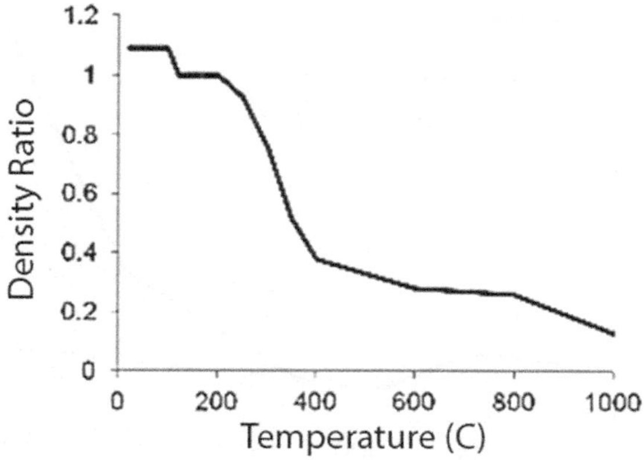

Figure 5-20. Density ratio for wood with a moisture content of 9% by weight. Source: Eurocode 5, CEN (2004), reprinted with permission.

Aside from temperature, the thermal conductivity and specific heat of uncharred wood are generally dependent on the density, moisture content, and grain orientation. The design values for these thermal properties provided in Eurocode 5 (CEN 2004) and in other sources are often effective properties and therefore have embedded assumptions that limit their range of applicability (König 2006). Thus, the designer should review these sources carefully prior to implementing the provided values.

The *Wood Handbook* (FPL 2010) provides the thermal conductivity k (in W/m·K) of solid wood at ambient temperature as

$$k = (0.0001941 + 0.000004064M)\rho_{od} + 0.01864 \tag{5-13}$$

The parameter M is the moisture content as a percent of weight and ρ_{od} is the oven-dried density (in kg/m^3). Values of conductivity are tabulated in the *Wood Handbook* for common wood types. This equation does not account for the increase in conductivity that occurs with temperature, which is typically 2–3% per 10°C (FPL 2010).

Eurocode 5 provides temperature-dependent thermal conductivity that accounts for the increased conductivity of the char layer (CEN 2004), which is shown in Table 5-6 and plotted in Figure 5-21. Note that the Eurocode model does not directly account for the effects of density or moisture content, and it uses effective properties rather than measured values to account for observed increases in values of properties at temperatures above 500°C. For most applications, the consideration of the upper-limit thermal conductivity would yield conservative results.

The specific heat of uncharred wood is dependent on temperature and moisture content, but it is not highly dependent on the species or density

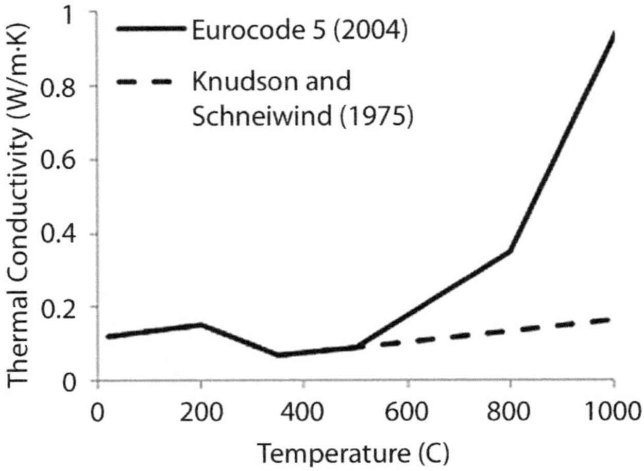

Figure 5-21. Conductivity of wood with temperature.
Source: Eurocode 5, CEN (2004).

(FPL 2010). The *Wood Handbook* (FPL 2010) provides the specific heat c (in kJ/kg·K) of oven-dried wood as a function of temperature T (in Kelvin) as

$$c = 0.1031 + 0.003867T \tag{5-14}$$

To account for moisture content M as a percent weight, the specific heat c (in kJ/kg·K) is given as

$$c = (c_{od} + c_w M)/(1 + M) + A_c \tag{5-15}$$

In Equation 5-15, c_{od} is the heat capacity of oven dry wood, c_w is the heat capacity of water (4.18 kJ/kg·K) and A_c is an adjustment factor to account for additional energy in the wood-water bond, which is given as

$$A_c = M(-6.191 + 2.36 \times 10^{-2}T - 1.33 \times 10^{-2}M) \tag{5-16}$$

The parameter T is the temperature in Kelvin. Note that these relationships are only valid for temperatures in the range of 44°F to 296°F (280 K to 420 K).

The specific heat of uncharred wood is also provided in the Eurocode 5, as shown in Table 5-6 and plotted in Figure 5-22 (CEN 2004). It should be noted that the Eurocode 5 model does not explicitly account for the dependence of specific heat on the moisture content. Instead, the model incorporates a sharp peak in specific heat capacity around 100°C to represent the energy absorbed owing to moisture evaporation, which is not included in the *Wood Handbook* model (FPL 2010). In addition, the Eurocode 5 model attempts to include the heat capacity of the char layer, allowing temperatures in excess of 150°C, whereas the thermal properties of the char layer are not

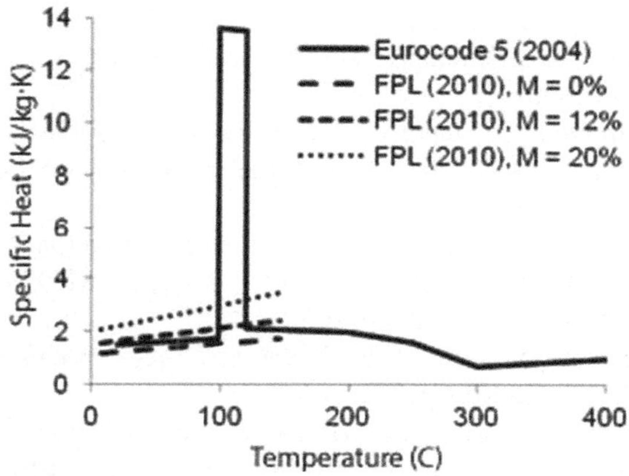

Figure 5-22. Specific heat capacity of wood at elevated temperature. Source: Eurocode 5, CEN (2004).

specified in the *Wood Handbook*. The designer should understand the limitations of these models prior to implementation. For most applications, the consideration of the lower limit specific heat would yield conservative results.

The emissivity of wood is fairly consistent across wood species once the surface has charred. The surface emissivity can therefore be taken as 0.90. For species-specific emissivity, refer to the *Wood Handbook* (FPL 2010).

REFERENCES

ACI/TMS (American Concrete Institute/The Masonry Society). 2007. *Code requirements for determining fire resistance of concrete and masonry construction assemblies*. ACI 216.1-07/TMS-216-07. Farmington Hills, MI: SCI; Longmont, CA: TMS.

Antal, M., and G. Varhegyi. 1995. "Cellulose pyrolysis kinetics: The current state of knowledge." *Ind. Eng. Chem. Res.* 34(3), 703–717.

ASCE/SEI (Structural Engineering Institute). 2005. *Standard calculation methods for structural fire protection*. ASCE/SEI/SFPE 29-05. Reston, VA: ASCE.

ASCE. 1992. *Structural fire protection, Manuals of practice (MOP) No. 78*, ASCE Fire Protection Committee. Reston, VA: ASCE.

ASTM International. 2008. *Standard test method for compositional analysis by thermogravimetry*. ASTM E1131. West Conshohocken, PA: ASTM International.

ASTM. 2009. *Standard test method for thermal conductivity of refractories by hot wire (platinum resistance thermometer technique)*. ASTM C1113. West Conshohocken, PA: ASTM International.

ASTM. 2011a. *Standard test method for linear thermal expansion of solid materials with a push-rod dilatometer*. ASTM E228. West Conshohocken, PA: ASTM International.

ASTM. 2011b. *Standard test method for determining specific heat capacity by differential scanning calorimetry*. ASTM E1269. West Conshohocken, PA: ASTM International.

ASTM. 2014. *Standard test method for determination of thermal conductivity of soil and soft rock by thermal needle probe procedure*. ASTM D5334. West Conshohocken, PA: ASTM International.

AWC (American Wood Council). 2012. *National design specification (NDS) for wood construction*. Leesburg, VA: AWC.

Babrauskas, V. 2001. "Ignition of wood—A review of the state of the art," *Interflam 2001*, Interscience Communications Ltd., London, 71–88.

Bailey, C. 2016. "Masonry thermal properties." Accessed May 9, 2018. http://www.mace.manchester.ac.uk/project/research/structures/strucfire/materialInFire/Masonry/thermalProperties.htm.

Benichou, N., and M. Sultan. 2000. "Fire resistance performance of lightweight wood-framed assemblies." *Fire Technol.* 36(3), 184–219.

Bentz, D., and K. Prasad. 2007. "Thermal performance of fire resistive materials: Characterization with respect to thermal performance models." *NISTIR 7401*. Gaithersburg, MD: NIST.

Bentz, D., C. White, K. Prasad, D. Flynn, D. Hunston, and K. T. Tan. 2009. "A materials science-based approach to characterizing fire resistive materials." *J. ASTM Int.* 6(5), 1–9.

Buchanan, A., and A. Abu. 2017. *Structural design for fire safety*, 2nd Ed. Hoboken, NJ: Wiley.

Burg, R., and B. Ost. 1994. "Engineering properties of commercially available high-strength concretes (including three-year data)." *Research and Development Bulletin RD104*. Skokie, IL: Portland Cement Association.

Butler, K., H. R. Baum, and T. Kashiwagi. 1995. "Heat transfer in an intumescent material using a three-dimensional Lagrangian model." *Proc., Int. Conf. Fire Research and Engineering*. Gaithersburg, MD: NIST.

Butler, K. 1997. "Physical modeling of intumescent fire retardant polymers." Chapter 15, *Polymeric foams*. Washington, DC: American Chemical Society.

Carino, N. J., M. A. Starnes, J. L. Gross, J. C. Yang, S. R. Kukuck, K. R. Prasad, et al. 2005. "Passive fire protection. Federal building and fire safety investigation of the World Trade Center disaster." *NIST NCSTAR 1-6A*. Gaithersburg, MD: NIST.

CEN (European Committee for Standardization). 2003. *Eurocode 6—Design of masonry structures. Part 1–2: General rules—Structural fire design*. EN 1996-1-2. Brussels: CEN.

CEN. 2004. *Eurocode 5—Design of timber structures. Part 1–2: General rules—Structural fire design*. EN 1995-1-2. Brussels: CEN.

CEN. 2005. *Eurocode 2—Design of concrete structures. Part 1–2: General rules—Structural fire design*. EN 1992-1-2. Brussels: CEN.

CEN. 2010. *Eurocode 3—Design of steel structures. Part 1–2: General rules—Structural fire design*. EN 1993-1-2. Brussels: CEN.

Di Blasi, C. 1993. "Modelling and simulation of combustion processes of charring and non-charring solid fuels." *Prog. Energy Combust. Sci.* 19, 71–104.

FABIG (Fire and Blast Information Group). 1993. "Fire resistant design of offshore topside structures." *Tech. Note No. 1*. Berkshire, UK: Steel Construction Institute.

FPL (Forest Products Laboratory). 2010. *Wood handbook: Wood as an engineering material. FPL-GTR-190*. Madison, WI: USDA Forest Service.

Harmathy, T. 1961. "A treatise on theoretical fire testing." *Symp. on fire test methods, ASTM STP301*. West Conshohocken, PA: ASTM International.

Harmathy, T. 1993. *Fire safety design and concrete*. Hoboken, NJ: Wiley.

Kodur, V., M. Dwaikat, and R. Fike. 2010. "High-temperature properties of steel for fire resistance modeling of structures." *J. Mater. Civ. Eng.* 22(5), 423–434.

Kodur, V., and T. Harmathy. 2016. "Properties of building materials," In *SFPE Handbook of Fire Protection Engineering*, 5th Ed. Gaithersburg, MD: Society of Fire Protection Engineers.

König, J. 2006. "Effective thermal actions and thermal properties of timber members in natural fires." *Fire Mater.* 30(1), 51–63.

Ljunggren, P. 1960. "Determination of mineralogical transformations of gypsum by differential thermal analysis." *J. Am. Ceram. Soc.* 43(4), 227.

Mesquita, L., P. A. G. Piloto, M. A. P. Vaz, and T. M. G. Pinto. 2009. "Decomposition of intumescent coatings: Comparison between a numerical method and experimental results." *Acta Polytech.* 49(1), 60–65.

Milke, J., S. Pessiki, and L. Phan. 2010. "Design of concrete structures," In *Best practice guidelines for structural fire resistance design of concrete and steel buildings, NIST Technical Note 1681*. Gaithersburg, MD: NIST.

Moghtaderi, B., V. Novozhilov, D. Fletcher, and J. H. Kent. 1997. "An integral model for the transient pyrolysis of solid materials." *Fire Mater.* 21(1), 7–16.

Moghtaderi, B. 2006. "The state-of-the-art in pyrolysis modelling of lignocellulosic solid fuels." *Fire Mater.* 30, 1–34.

Opstad, G. 2010. "Thermal properties of intumescent passive fire protection materials." *Proc., Int. Conf. Safety and Environment in Process Industry*. Florence, Italy: *Chemical Engineering Transactions* 19, 237–242 .

Purkiss J. 1996. *Fire safety engineering design of structures*. Milton Park, Abingdon, UK: Taylor & Francis.

Schaffer, E. 1967. "Charring rate of selected woods–Transverse to grain." *Res. Pap. FPL 69*. Madison, WI: USDA Forest Service, Forest Products Laboratory.

Schaffer, E. 1984. "Structural fire design: Wood." *Forest Products Laboratory research paper FPL 450*. Madison, WI: USDA Forest Service.

Spearpoint, M., and J. Quintiere. 2000. "Predicting the burning of wood using an integral model." *Combust. Flame* 123(3), 308–325.

Sultan, M. 1996. "A model for predicting heat transfer through non-insulated unloaded steel-stud gypsum board wall assemblies exposed to fire." *Fire Technol.* 32(3), 239–259.

Szoke, S. 2006. "Resistance to fire and high temperatures." In *STP169D Significance of tests and properties of concrete and concrete-making materials*. West Conshohocken, PA: ASTM International.

Van Geem, M., J. Gajda, and K. Dombrowski. 1996. "Thermal properties of commercially available high-strength concretes." *Cem. Concr. Aggregates* 19(1), 38–54.

Wang, Y., I. Burgess, F. Wald, and M. Gillie, M. 2013. *Performance-based fire engineering of structures*. Boca Raton, FL: CRC Press.

West, R., and W. Sutton. 1954. "Thermography of gypsum." *J. Am. Ceram. Soc.* 37(5), 221–224.

White, R. 1988. "Charring rates of different wood species." Ph.D. dissertation, Madison, WI: University of Wisconsin.

White, R. 2016. "Analytical methods for determining fire resistance of timber members." In *SFPE handbook of fire protection engineering*, 5th Ed. Quincy, MA: National Fire Protection Association; Bethesda, MD: Society of Fire Protection Engineers.

CHAPTER 6
HEAT TRANSFER CALCULATION

The thermal response of structural elements should be determined considering the three modes of heat transfer: conduction, convection, and radiation. The thermal properties of applied insulation (e.g., specific heat, thermal conductivity, density, and emissivity) should also be considered (see Chapter 5). The SFPE S.02 standard provides requirements for heat transfer calculations to predict the thermal response of structural elements to time-dependent thermal boundary conditions imposed by fires (SFPE 2015). The designer should follow this standard or similar industry standards for such calculations.

All three modes of heat transfer can be expected to occur when a structural system is heated by fire. Conduction is heat transfer through a solid. Conduction occurs from one element to another that are in physical contact with each other and have a temperature difference. For instance, there would be conduction heat transfer from the insulating material (e.g., an SFRM) to the underlying structural element during fire exposure. Conduction heat transfer follows Fourier's equation, which is given in Equation 6-1.

$$q = -kA\frac{\partial T}{\partial x} \tag{6-1}$$

where

q = rate of heat transfer,

k = thermal conductivity (which is a property of the material), and

$A\frac{\partial T}{\partial x}$ = temperature gradient in the direction of heat flow.

Convection involves the transfer of heat by a circulating fluid. Convection is the method by which heat is transferred from hot gases created by a fire to the portions of the structure with which they come into contact. Convection heat transfer follows Newton's law of cooling, which follows Equation 6-2.

$$q = hA\Delta T \tag{6-2}$$

where
q = rate of heat transfer,
h = convection heat transfer coefficient,
A = exposed surface area of the object, and
ΔT = difference in temperature between the exposed object and the fluid with which it is in contact.

A convective heat transfer coefficient of 25 W/m²K and 50 W/m²K is recommended for a standard fire and hydrocarbon fire, respectively (SFPE 2015, CEN 2005). For actual fire conditions, a convective heat transfer coefficient of 35 W/m²K is recommended (CEN 2005).

Radiation heat transfer involves the transfer of energy via electromagnetic waves. These electromagnetic waves travel at the speed of light and via line of sight. All objects emit thermal radiation in proportion to their temperature raised to the fourth power. The net change in energy from radiation heat transfer is equal to the difference between the amount of energy radiated from an object and that which it receives from other objects. Radiation heat transfer between two objects follows Equation (6-3).

$$q = \varepsilon F_G \sigma A\left(T_1^4 - T_2^4\right) \tag{6-3}$$

where
q = rate of heat transfer,
ε = emissivity of the object,
F_G = geometric view factor between the two objects,
A = exposed surface area of the object,
σ = Stefan–Boltzmann constant, and
T_1 and T_2 = temperatures of the objects.

The emissivity of an object must be between 0 and 1.0. The emissivity value should be based upon the material that is exposed to the fire (e.g., gypsum board). A conservative estimate of the emissivity is 1.0, which represents an ideal blackbody. The geometric view factor is a measure of how much of the energy that is radiated by one object is "seen" by the other. Tables of view factors can be found in heat transfer textbooks for commonly shaped objects. The view factor must be between zero and one. If an object is surrounded by another (e.g., a column in a room), the view factor is equal to 1.0.

For SFED calculations, *resultant* emissivity values are often cited in the literature for radiation exchange between a structural element and an enclosure fire. The resultant emissivity ε is given by

$$\varepsilon = \frac{1}{\frac{1}{\varepsilon_e} + \frac{1}{\varepsilon_r} - 1} \tag{6-4}$$

The parameter ε_e is the emissivity of the emitting surface, and ε_r is the emissivity of the receiving surface (Buchanan and Abu 2017). Parkinson et al. (2009) provided resultant emissivity values for steel columns and beams based on their configuration with respect to fire. Caution should be exercised when using resultant emissivity values. In particular, the equation for calculating the resultant emissivity is derived from the assumption that the structure and the fire are infinitely parallel gray surfaces that transmit radiation through a perfectly transparent medium. This assumption is most appropriate for heat transfer in a furnace where combustion is highly efficient, but it may be inadequate for natural fire exposures in which combustion products will contribute to the radiation exchange (Ghojel and Wong 2005). In addition, the resultant emissivity is commonly assumed to be constant, whereas it is generally dependent on temperature (Mooney 1992). Radiation between surfaces and emitting/absorbing gases is very complex. If the emissivity of the gas is negligible (e.g., furnace condition) and the walls have low density and thermal conductivity, this assumption is reasonable, granted that the fire is fully developed and the soot volume fraction is very high. Any intermediate condition should be further analyzed by the designer, considering the partially absorbing/emitting medium.

The SFPE S.02 standard recommends that the emissivity of radiating fire gases be taken as 1.0 or some value just less than 1.0. Thus, the incident radiation is calculated as the blackbody radiation temperature equal to the gas temperature (SFPE 2015).

6.1 STEEL STRUCTURES

Because of its high thermal conductivity, steel exhibits a relatively homogeneous temperature distribution when subjected to uniform heating. For design purposes, the thermal analysis of unprotected steel structures can generally be accomplished with a lumped mass heat transfer model. When nonuniform heating must be accounted for, a conduction heat transfer analysis can be performed using the finite element or finite difference method. The thermal analysis of protected steel members requires consideration of any applied insulation, such as SFRM. Issues regarding the variability of SFRMs and the effects of insulation damage on the thermal response of steel structures are discussed in the following sections.

6.1.1 Variability of SFRM Thickness and Density

The application of SFRM to structural steel members has inherent variability in terms of the thickness and density over surfaces. The mixture of dry components with water during preparation and subsequent curing controls the density of the SFRM. The uniformity and consistency of spray application governs the variability of SFRM thickness on structural steel members.

NIST investigated the sensitivity of steel temperature rise from fire exposure to the variability in SFRM thickness. The effect of variability in thickness was modeled by imposing a normal probability distribution on the SFRM thickness along the length of a steel plate. Based on finite element model analyses, it was shown that when the SFRM thickness is variable, the isotherms in the steel follow the shape of the SFRM surface contour. Thus, the temperature history at any point in the steel strongly depends on the local thickness of the insulation. Moreover, it was found that increased variation in thickness reduced the *effective thickness* (i.e., equivalent uniform thickness) of the SFRM as it pertains to the thermal response of the steel (Carino et al. 2005). Hence, the designer should reduce the effective insulation thickness for thermal response analyses to account for inherent variability in the in-situ thickness.

The ASTM E605 standard (ASTM 2006) specifies testing procedures to determine the thickness and density of the SFRM and whether an installation is deficient or not in this respect. For a given beam, joist, or column, an average of 18, 14, and 24 thickness measurements, respectively, are required to be recorded. The structural member is deficient if an individual measured thickness is more than ¼ in. less or more than 25% less than the required design thickness. Hence, a reduction in the effective SFRM thickness of ¼ in. would be conservative.

For an SFRM installation, ASTM E605 (ASTM 2006) also requires that one density test on each of the following protected structural elements be conducted at random per floor or per every 10,000 sq. ft., whichever provides the greatest number of tests: the flat portion of a deck; a beam, either the bottom of the beam lower flange or the beam web; and a column, either the column web or the outside of one of the column flanges. When density falls below the minimum average and minimum individual values per the manufacturer's specifications, a similar, randomly selected element in the same area shall be tested. If the average density of the two elements is then met, the area passes. If the average is not met, then those elements in that area must be corrected.

Based upon the construction requirements described in the ASTM E605 standard (ASTM 2006), the designer should consider that the SFRM is no more than ¼ in. less than the thickness specified. However, since SFRM overspray is not penalized under this standard, the designer should also be aware that the SFRM may be significantly thicker than specified, which is conservative for SFED. Also, the designer should consider the variability of SFRM density as ±5%.

6.1.2 Localized Insulation Damage

The loss of applied insulation from a structural steel member can adversely affect its performance under fire exposure. Conventional-type SFRM coverage often is imperfect because it can be removed easily from steel, and poor application, natural deterioration, or other effects may cause localized deficiencies. Moreover, standard fire testing does not capture the influences of these important in-situ conditions.

Using finite element modeling, NIST investigated the effect of localized loss of SFRM on the thermal response of a steel plate to fire. It was found that the bare steel at the missing insulation region approached the gas temperature quickly, but more important, the gap in the insulation led to transmission of heat throughout the thickness of the steel plate (Carino et al. 2005). Ryder et al. (2002) analyzed the effect of the loss of SFRM on the thermal response of steel columns. In the study, finite element modeling was used to estimate the temperature rise within a steel column with varying sections of SFRM removed. To a certain extent, it was found that a column with a higher thickness-to-depth ratio would better withstand the removal of SFRM owing to its increased thermal capacity (Ryder et al. 2002). LaMalva (2010) performed finite element analyses of typical steel framing with varying extents and locations of SFRM damage. It was found that in certain cases the location of SFRM damage proved to be more important than the extent of the SFRM damage as it pertains to overall structural response (e.g., SFRM damage at the girder to column interface).

Wang and Li (2009) conducted fire testing and analyses of steel columns with partial loss of SFRM. The steel columns had an axial load applied that was kept constant with a load ratio of 0.55, and slenderness values of 47.2 and 119.6 in the strong and weak axes, respectively. The damaged length of SFRM was 7% of the column length for one specimen and 14% for the other. These experimental and analytical results demonstrate that the SFRM damage length has a significant effect on the thermal response of steel columns and that the failure of the specimens mainly resulted from the buckling or yielding at the section where SFRM was missing. Given the slenderness and applied load ratio of the tested steel columns, it was found that when the SFRM damage was dimensionally long, failure of the column was due to buckling, and when the damage was dimensionally short, failure was due to yielding (Wang and Li 2009).

Braxtan and Pessiki (2011) analyzed damage patterns in SFRM on steel moment frame beam–column assemblies owing to a strong seismic event and the thermal consequences of this damage when exposed to a post-earthquake fire (see Section 4.2.5). At story drifts of 3% to 4%, SFRM damage was concentrated in the beam flanges where large inelastic deformation and local buckling occurred. The results of a subsequent heat transfer analysis demonstrated that various SFRM damage patterns on the beams adjacent to the column caused an increase in temperature of the column along varying percentages of its height as compared to the undamaged case.

Under normal conditions, it is reasonable for a designer to assume that applied insulation will not be damaged. However, in areas of high traffic or harsh environmental conditions, the designer should account for the possibility of damage due to impact and/or erosion. This may involve specification of higher-strength SFRM and/or mesh reinforcement or by explicit consideration during a thermal response analysis. If post-hazard event fire exposures are being considered per a specific discretionary performance objective (see Section 3.2), the designer should assume an appropriate level of damage to the applied insulation (e.g., at areas of inelastic deformation). This would involve comparing the bond strength of the applied insulation to the stresses that would occur at the insulation–structure interface, which may require specific information from the insulation manufacturer.

6.1.3 Thermal Bridging

Because of their relatively high thermal conductivity, steel structures are particularly susceptible to thermal bridging effects, in which heat is transferred across certain interfaces through the steel. Thermal bridging is commonly related to energy conservation issues (e.g., building envelope design), but it also can be an important consideration for SFED. For instance, thermal bridging can occur at the junctions of protected and unprotected steel under fire exposure, such as unprotected secondary framing into protected columns. In this case, the unwanted heat transfer would increase the heating of the column locally.

Thermal bridging effects can be mitigated by use of additional insulation. For instance, a secondary steel member that frames into a column may be protected with insulation for a segment of its length (e.g., several feet) at the side that interfaces with the column. Less commonly, alternative materials that have lower thermal conductivity than carbon steel may be used at critical connections (e.g., stainless steel) or at their interfaces (e.g., fiber-reinforced polymer) (Anderson et al. 2012). In all cases, the designer should consider possible thermal bridging effects as part of any heat transfer analysis. However, it is relatively unlikely that these effects would govern an SFED.

6.2 CONCRETE STRUCTURES

To evaluate the thermal response of concrete members under fire exposure, a conduction heat transfer analysis must be performed using the finite element or finite difference method. In such cases, the designer may ignore the reinforcing steel in the heat transfer analysis when it has a comparatively small mass. In these cases, it is sufficient for the designer to consider the reinforcement temperature as that of the modeled concrete at the location of the reinforcement. When consideration of the reinforcing steel is needed, the thermal properties of structural steel may be used (NIST 2010). When large voids exist within concrete members (e.g., hollowcore precast concrete), the cavity radiation within these voids should be explicitly accounted for.

Moisture evaporation has a noticeable effect on the thermal response of concrete, but it can be conservatively ignored for design purposes (CEN 2005). However, the loss of concrete cover due to spalling should be taken into account if spalling is expected to occur. The prediction of fire-induced spalling is discussed in Chapter 8.

The ACI 216.1 standard provides temperature contours for regular slabs, walls, and rectangular and angular members under standard fire exposure for simplified fire resistance calculations (ACI/TMS 2007). However, this information should not be relied upon for SFED.

6.3 TIMBER STRUCTURES

Three calculation methods have been approved by U.S. building codes for determining the fire resistance ratings of timber structures using SFRD: the T. T. Lie method (AWC 2000), the component additive method (AWC 2010), and a mechanics-based method contained in the NDS (AWC 2012). Within SFRD, these methods allow the fire resistance ratings of wood assemblies to be determined without the need for extensive heat transfer analyses. These methods are generally not applicable to SFED because they are limited to the standard fire exposure. However, the mechanics-based method can be adjusted for consideration of structural design fires if the nonlinear char rate associated with the fire exposure is known.

Thermal analyses of wood members under fire exposure must account for thermal degradation (i.e., pyrolysis) of the wood at temperatures above the ignition temperature. This can be accomplished using a simplified model based on a prescribed charring rate or by including a thermal decomposition model that enforces conservation of energy and mass at the char layer (see Section 5.7). However, the designer should understand that simplified char models typically only apply to timber sections 5 in. × 5 in. or larger and that the use of such models for dimensional lumber may yield unconservative results.

6.4 NONUNIFORM HEATING

Thermal response analyses should consider the presence of thermal gradients in structures subjected to nonuniform heating. During fire exposure, a temperature gradient may develop through the depth/thickness of structural elements. For example, a temperature gradient through the depth of steel beams below concrete floors usually develops under fire exposure because the floor slab acts as a heat sink for the steel beam below (Figure 6-1). Also, perimeter columns may experience nonuniform heating during fire exposure owing to parts of the column's perimeter being shielded from the fire.

During fire exposure, thermal gradients may develop across the longitudinal axis of structural elements owing to heat dissipation effects. For example, a steel beam connection in close proximity to a protected steel column may experience significantly lower temperatures than the beam mid-span during fire exposure (Figure 6-2). This would be because of heat dissipation from the steel connection to the protected column, which may have more insulation and/or direct connection to cooler structural elements residing in compartments without fire exposure. This effect is not realistically captured in standard fire testing and can have an appreciable effect on the performance of the connection under heating (LaMalva 2011). In addition, beams subjected to localized fires may experience significant thermal gradients in the longitudinal direction because of spatially varying surface heat fluxes.

Figure 6-1. Temperature gradient through steel beam depth (2D thermal model). Source: Kevin LaMalva.

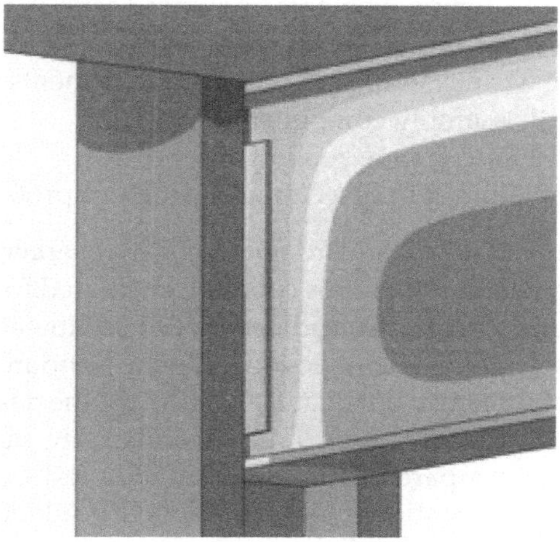

Figure 6-2. Heat dissipation from a beam connection to column (3D thermal model). Source: Kevin LaMalva.

The designer should consider any relationships between the thermal and mechanical responses of structures to determine whether nonuniform heating can be conservatively ignored or should be explicitly accounted for in the thermal analysis (see Section 7.2.2). For instance, a column that has a significant temperature gradient over the depth will undergo significant thermal bowing, resulting in geometrically nonlinear effects that may significantly reduce the structural capacity. Structural response under nonuniform heating is further discussed in the literature (Culver 1972; Ossenbruggen et al. 1973; Kruppa 1981; Witteveen and Twilt 1981; Becker 2002).

6.5 MECHANICAL INTEGRITY

Heat transfer calculations inherently assume that the applied insulating materials will remain in place during the fire exposure. Determination of adequate mechanical integrity in the installed condition is most typically evaluated using standard fire testing within SFRD. This provides documentation that the insulation maintains adequate mechanical integrity under the furnace exposure. Mechanical integrity performance includes but is not limited to the following:

- The adhesion and cohesion of SFRM [e.g., field-qualified per ASTM E736 (2015)];
- Degradation of nonstructural assembly components such as gypsum encasement;

- Adequacy of fastening elements, including screws, nails, and brackets used to secure protective membranes;
- Adequacy of wiring, strapping, and similar elements used to suspend nonstructural assembly components;
- Susceptibility for concrete spalling; and
- Adequacy of joints at intersections of different protection systems.

Information derived from standard fire testing as it pertains to the mechanical integrity of insulating materials could be evaluated for applicability to structural design fire scenarios considered for SFED. However, the mechanical integrity of materials or assemblies qualified by standard fire testing may not automatically be deemed adequate. Accordingly, the designer must exercise caution if a structural design fire is significantly more severe (e.g., sharper growth) as compared to the standard fire test exposure. Overall, the designer should be confident that the applied insulation materials will stay in place long enough to fulfill the required performance objectives. This may require the specification of applied insulation that has been qualified for challenging exposures [e.g., the UL 1709 exposure (UL 2011)]. Also, the designer may consult the relevant manufacturer for specific information or reference specific test data. At a minimum, the designer should adhere to the manufacturer's installation requirements (e.g., restrictions on the use of steel primers prior to the application of SFRM). The industry has yet to develop a test method to qualify the mechanical integrity performance of applied insulation for nonstandard fire exposure and generalized application in SFED. Hence, it is currently within the purview of the designer to provide sufficient evidence, analysis, and judgment in this respect. This may require specific information from the relevant manufacturer.

Large deflection of long-span beams under fire exposure may cause applied insulation to undergo adhesion failure or gypsum board encasement systems to become unstable (Carino et al. 2005). Unfortunately, geometric limitations of test furnaces only allow for beams to be tested at relatively short spans, which are typically a poor reflection of actual building construction. The designer should consider these aspects accordingly for a SFED. Again, this may require the designer to consult the relevant manufacturer for specific information or reference specific test data.

SFRD listings define appropriate extrapolations from test conditions, such as application to heavier steel sections than tested. However, dimensional scaling is usually not appropriate to qualify the mechanical integrity of applied insulation as part of SFED. For instance, the endothermic calcination process that SFRM undergoes upon heating (which enhances its insulating capability) cannot be scaled based on its thickness. Other examples include a very thin material that may not maintain mechanical integrity or an intumescent material that will yield little or no additional benefit if a thicker layer is applied. Hence, the use of applied insulation thicknesses for

SFED that are outside the range included in specific SFRD listings should not be done without validation testing or specific information from the manufacturer, as materials are often not as useful at thicknesses outside the range included in the testing. In this respect, the availability of information from relevant manufacturers is paramount and perhaps indicative of their competitive advantage for inclusion in SFED applications.

6.6 METHODS OF HEAT TRANSFER ANALYSIS

Heat transfer analyses are used by designers to estimate the temperature field of the structural system based on the imposed thermal boundary conditions by fire exposure. The nature and level of complexity of the heat transfer model is dependent on the physics that are being captured in the analysis. Simplifying assumptions should be justified through requisite verification and validation exercises, as described in the SFPE S.02 standard (SFPE 2015). Lumped mass, one-dimensional (1D), and two-dimensional (2D) heat transfer analyses are most commonly used for SFED, although there are instances in which three-dimensional (3D) heat transfer analyses may be justified (e.g., the presence of strong heat sinks in multiple directions). These thermal analysis techniques are introduced below.

6.6.1 Lumped Mass Method

The lumped mass method is a step-by-step (i.e., quasi-steady) energy balance calculation technique that assumes a uniform temperature of the cross section at any point in time, as shown in Equation (6-5). This method is most commonly applied to unprotected structural steel sections because the high thermal conductivity of steel lends itself well to the primary assumption of the calculation method. It is reasonable to use this simplified method in cases where the temperatures from fire are relatively uniform throughout the compartment (Buchanan 2017) and longitudinal heat sink effects are not significant (LaMalva 2011).

$$\Delta T_s = \frac{F}{V}\frac{1}{\rho_s c_s}\left\{h_c\left(T_f - T_s\right) + \sigma\varepsilon\left(T_f^4 - T_s^4\right)\right\}\Delta t \qquad (6\text{-}5)$$

where
ΔT_s = change in steel temperature over the time step,
Δt = time step,
F = the surface area per unit length of the member,
V = volume of steel per unit length of the member,
ρ_s = density of steel,
c_s = specific heat of steel,

h_c = convective heat transfer coefficient,
T_f = temperature of the fire environment, and
T_s = steel temperature.

Before applying the lumped mass method, the designer should consider the applicability of the method. The Biot number is defined as the ratio of the resistance of internal heat flow to the resistance of external heat flow. For a flat plate, the Biot number is the combined heat transfer coefficient at the surface multiplied by the plate thickness, and this term is then divided by the thermal conductivity of the plate. Theoretical considerations suggest that the lumped mass method is a reasonable approach if the Biot number is less than about 10% or 0.1 (Kreith 1965).

For structural members protected with insulating materials or those with temperature-dependent properties, a finite element or finite difference model should be used for calculating the temperatures (Buchanan and Abu 2017). If the lumped mass method is deemed suitable for analysis of a protected member, the calculation procedure is similar to that for unprotected members except the equation is adjusted to account for the insulation. It is assumed that the temperature gradient in the insulation is linear. For a steel member protected with insulation, this approximation improves with decreasing insulation thickness.

The rate of temperature rise of a protected or unprotected structural steel member exposed to fire primarily depends on the section factor, which is a measure of the ratio of the heated perimeter to the area of the cross section. The lumped mass method accounts for this important parameter because the rate of heat input is directly proportional to the area exposed to the fire environment, and the subsequent rate of temperature increase is inversely proportional to the heat capacity of the member (equal to the product of the specific heat, the density, and the volume of the steel segment) (Buchanan and Abu 2017).

The lumped mass method can account for the temperature dependence of the insulation's heat capacity if this parameter is updated at each increment. Otherwise, this method may not be appropriate for insulating materials that exhibit a strong dependence of its heat capacity on temperature (e.g., gypsum-based material that undergoes an endothermic calcination process during heating). If longitudinal heat sink effects are significant (e.g., steel girder connection to a heavy column), the lumped mass method may only provide information concerning the heating of the section at a significant distance away from heat sinks (e.g., at the girder mid-span). Moreover, the vertical heat sink effect of a concrete slab would not be accounted for using this method, possibly leading to a significant overestimation of the section's temperature history.

Although the assumptions of the lumped mass method lend themselves well to steel members, this method would not be appropriate for sections with significant thermal resistance such as a concrete member. Notwithstanding, a robust steel member with thick flanges and web may develop appreciable thermal gradients within its cross section, which would not be accounted for using the lumped mass method. Typically, the intersection of the web with the flanges remains at a lower temperature than other parts of the cross section (e.g., web mid-height). The designer should evaluate whether these behaviors would significantly affect the structural response.

6.6.2 Finite Element Method

The finite element method is a numerical method that is used to solve field problems, including conduction heat transfer in solids. The method involves discretizing the computational domain into a network of elements and nodes such that temperatures are calculated at discrete points in the continuum. To derive the finite element formulation, the weak form of the governing conservation equation is expressed over a single element's domain. The field variable (e.g., temperature, in the case of heat transfer analysis) is approximated over the element's domain as a function of the nodal values using interpolation functions or shape functions. Through a series of manipulations, a set of element equations of the following form is derived and assembled for all elements in the domain:

$$[c]\{\dot{T}\}+[k]\{T\}=\{r\} \tag{6-6}$$

where
$[c]$ = heat capacity matrix,
$[k]$ = conductivity matrix,
$\{r\}$ = vector of thermal loads,
$\{T\}$ = vector containing the nodal temperatures, and
$\{\dot{T}\}$ = the partial derivatives of temperature with time.

By discretizing the domain in space and time, the governing differential equations are approximated by a set of algebraic equations that can be solved computationally. For more comprehensive information about the finite element method, the designer should review Hutton (2013) or similar texts. Figure 6-3 shows the results of a 2D finite element thermal model of a steel column protected with SFRM at three instances in time.

Many commercial software programs have the capability to solve conduction heat transfer problems, and therefore they can be applied to thermal analyses for SFED. In addition, several special-purpose finite element codes have been developed specifically for SFED (Iding et al. 1996, Wickström and Pålsson 1999, Franssen 2005). Special-purpose codes have features such as material models already embedded in the software for convenience. When selecting a computer code for performing heat transfer analyses, it is

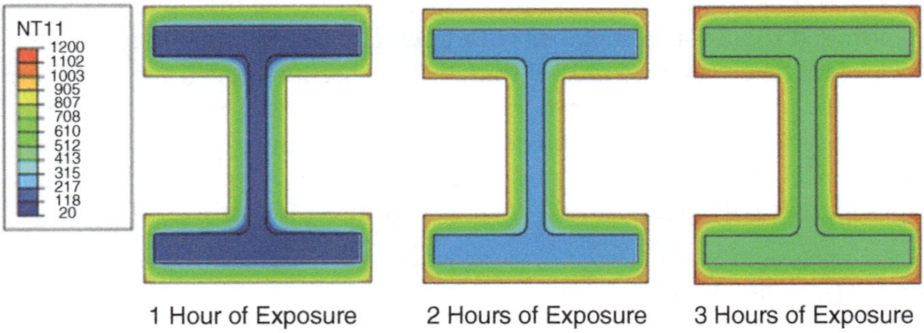

Figure 6-3. Two-dimensional thermal model results of steel column protected with SFRM.
Source: Kevin LaMalva.

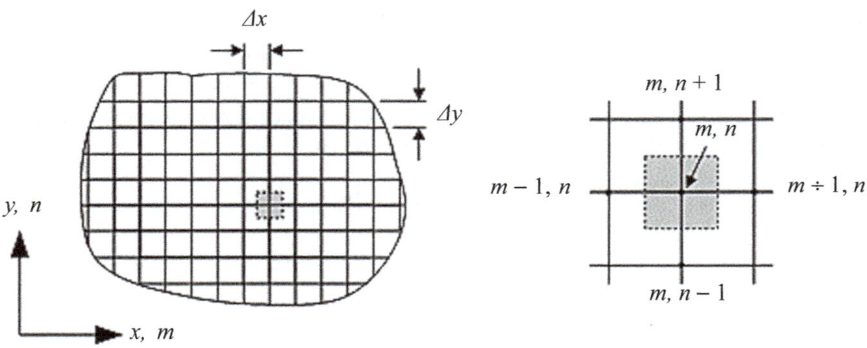

Figure 6-4. Two-dimensional finite difference mesh.
Source: Ann Jeffers.

important that the strengths and limitations of the software are understood. At a minimum, the validity of the calculation model and the accuracy of the material properties should be verified prior to using the code (Wickström 2016). Annex A of the SFPE S.02 standard provides verification cases that the designer may utilize for these purposes (SFPE 2015).

6.6.3 Finite Difference Method

The finite difference method is a numerical technique that is commonly used to solve the equations of heat transfer in solids. In the finite difference method, a network of nodes is generated over the computational domain such that temperatures are calculated at discrete points in the continuum, as illustrated in Figure 6-4. The mesh or grid that is selected is dependent on the geometric features of the continuum being modeled, as well as the degree of accuracy that is needed in the solution.

The governing equations of heat transfer are derived by applying principles of energy conservation. The temperature gradients that appear in the governing equations are replaced by finite difference approximations. For example, the following approximation is often substituted for the derivative of the temperature gradient in the diffusive term that appears in the following energy balance equation:

$$\left.\frac{\partial^2 T}{\partial x^2}\right|_{m,n} = \frac{T_{m+1,n} - 2T_{m,n} + T_{m-1,n}}{(\Delta x)^2} \tag{6-7}$$

The parameter $T_{i,j}$ is the temperature of node (i, j), and Δx is the spacing between nodes. The use of finite differences allows the differential equation over the domain to be replaced by a system of algebraic equations that can be solved numerically.

Finite difference calculations for heat transfer analysis are covered extensively in texts on heat transfer analysis. The method has been widely used to perform linear and nonlinear heat transfer calculations, although it is less commonly used in SFED practice. The primary limitation of the finite difference method is that it cannot easily handle complex geometries (in a reasonable computational time) as can be done using the finite element method.

REFERENCES

ACI/TMS (American Concrete Institute/The Masonry Society). 2007. *Code requirements for determining fire resistance of concrete and masonry construction assemblies.* ACI 216.1-07/TMS-216-07. Farmington Hills, MI: ACI; Longmont, CA: TMS.

Anderson, J., J. D'Aloisio, D. Delong, R. Miller-Johnson, K. Oberdorf, R. Ranieri, et al. 2012. "Thermal bridging solutions: Minimizing structural steel's impact on building envelope energy transfer." Supplement to *Modern Steel Construction*. Chicago: American Institute of Steel Construction.

ASTM International. 2006. *Standard test methods for thickness and density of sprayed fire-resistive material (SFRM) applied to structural members.* ASTM E605. West Conshohocken, PA: ASTM International.

ASTM. 2015. *Standard test method for cohesion/adhesion of sprayed fire-resistive materials applied to structural members.* ASTM E736. West Conshohocken, PA: ASTM International.

AWC (American Wood Council). 2000. *Design of fire-resistive exposed wood members.* DCA No. 2. Leesburg, VA: AWC.

AWC. 2010. *Component additive method (CAM) for calculating and demonstrating assembly fire endurance.* DCA No. 4. Leesburg, VA: AWC.

AWC. 2012. *National design specification (NDS) for wood construction*. Leesburg, VA: AWC.

Becker, R. 2002. "Structural behavior of simple steel structures with non-uniform longitudinal temperature distributions under fire conditions." *Fire Saf. J.* 37(5), 495–515.

Braxtan, N., and S. Pessiki. 2011. "Postearthquake fire performance of sprayed fire-resistive material on steel moment frames." *J. Struct. Eng.* 137(9), 946–953.

Buchanan, A., and A. Abu. 2017. *Structural design for fire safety*, 2nd Ed. New York: Wiley.

Carino, N., M. A. Starnes, J. L. Gross, J. C. Yang, S. R. Kukuck, K. R. Prasad, and R. W. Bukowski. 2005. "Passive fire protection, federal building and fire safety investigation of the World Trade Center disaster." *NIST NCSTAR 1-6A*. Gaithersburg, MD: NIST.

CEN (European Committee for Standardization). 1995. *Eurocode 1—Basis of design and actions on structures, Part 1–2: Actions on structures—Actions on structures exposed to fire*. ENV 1991-1-2. Brussels: CEN.

CEN. 2005. *Eurocode 2—Design of concrete structures, Part 1–2: General rules—Structural fire design*. EN 1992-1-2. Brussels: CEN.

Culver, C. 1972. "Steel column buckling under thermal gradients." *J. Struct. Div., ASCE*, 92(8), 1853–1865.

Franssen, J. 2005. "SAFIR: A thermal/structural program for modelling structures under fire." *Eng. J.* 42(3), 143–150.

Ghojel, J., and M. Wong. 2005. "Heat transfer model for unprotected steel members in a standard compartment fire with participating medium." *J. Construct. Steel Res.* 61(6), 825–833.

Hutton, D. 2013. *Fundamentals of finite element analysis*. Columbus, OH: McGraw-Hill.

Iding, R., B. Bresler, and Z. Nizamuddin. 1996. "FIRES-T3: A computer program for the fire response of structure-thermal (three-dimensional version)." *NIST GCR 95-682*. Gaithersburg, MD: NIST.

Kreith, F. 1965. *Principles of heat transfer*, 2nd Ed. Scranton, PA: International Textbook Co.

Kruppa, J. 1981. "Some results on the fire behavior of external steel columns." *Fire Saf. J.* 4(4), 247–257.

LaMalva, K. 2010. "Sensitivity analyses of structural steel connection fire performance." *Proc., SEI Structures Congress*. Reston, VA: ASCE.

LaMalva, K. 2011. "Thermal response of steel structures to fire: test versus field conditions." *J. Fire Protect. Eng.* 21(4), 285–299.

Mooney, J. 1992. "Surface radiant-energy balance for structural thermal analysis." *Fire Mater.* 16(2), 61–66.

NIST (National Institute of Standards and Technology). 2010. "Design of concrete structures." In *Best practice guidelines for structural fire resistance design of concrete and steel buildings, NIST Tech. Note 1681*, L. T. Phan, T. P. McAllister, J. L. Gross, and M. J. Hurley (eds.), Gaithersburg, MD: NIST.

Ossenbruggen, P., V. Aggarwal, and C. Culver. 1973. "Steel column failure under thermal gradients by member." *J. Struct. Div., ASCE*, 99(4), 727–739.

Parkinson, D., V. Kodur, and P. Sullivan. 2009. "Performance-based design of structural steel for fire conditions." *ASCE Manuals and Reports on Engineering Practice, Vol. 114*. Reston, VA: ASCE.

Ryder, N. L., S. D. Wolin, and J. Milke. 2002. "An investigation of the reduction in fire resistance of steel columns caused by loss of spray applied fire protection." *J. Fire Protect. Eng.* 12(1), 31–44.

SFPE (Society of Fire Protection Engineers). 2015. *Engineering standard on calculation methods to predict the thermal performance of structural and fire resistive assemblies*. SFPE S.02. Gaithersburg, MD: Society of Fire Protection Engineers.

UL (Underwriters Laboratories). 2011. *Standard for rapid rise fire tests of protection materials for structural steel.* UL 1709. UL: Northbrook, IL.

Wang, W., and G. Li. 2009. "Behavior of steel columns in a fire with partial damage to fire protection." *J. Constr. Steel Res.* 65(6), 1392–1400.

Wickström, U., and J. Pålsson. 1999. "Scheme for verification of computer codes for calculating temperature in fire-exposed structures." *SP REPORT 1999:36*. Borås, Sweden: SP Swedish Testing and Research Institute.

Wickström, U. 2016. "Methods for predicting temperatures in fire-exposed structures." *SFPE Handbook of fire protection engineering*, 5th Ed. Gaithersburg, MD: Society of Fire Protection Engineers.

Witteveen, J., and L. Twilt. 1981. "A critical view on the results of standard fire resistance tests on steel columns." *Fire Saf. J.* 4(4), 259–270.

PART 3

STRUCTURAL RESPONSE TO FIRE EFFECTS

CHAPTER 7
STRUCTURAL FIRE EFFECTS

Fire effects caused by structural design fires should be evaluated for an SFED. Fire effects include thermally induced forces and their related displacements, deflections, and degraded material properties. Temperature-dependent effects result in temperature-dependent structural resistance that may affect a structure's ability to support its gravity loads. If structural failure is avoided during the fire exposure, the residual post-fire state of the structural system during the cooling phase should also be evaluated for its stability and ability to meet the required performance objectives (e.g., the safety of occupant refuge areas). This chapter introduces concepts and considerations associated with fire effects on structures for structural analysis that are addressed in more detail in Chapters 8 and 9.

7.1 OVERVIEW

Perhaps the most significant effect of fire exposure on a structural system is the potential reduction in structural stability and load-bearing capacity, which require the inclusion of temperature-dependent material properties and thermal effects in an SFED. Structural members may lose strength and stiffness at elevated temperatures. Fire-exposed members may also undergo thermal expansion, the restraint of which can lead to additional induced internal forces. Some materials, such as wood and concrete, may also experience a reduction in cross section owing to charring or spalling.

Temperature gradients over a structural member cross section can cause differential thermal expansion between the hotter and cooler portions of the section. Differential expansion can result in curvature for pinned or fixed members. For members with partial or full rotational end restraint, a temperature gradient may result in a strain gradient through the depth of the member owing to restraint of thermal expansion.

Heated structural members may undergo deflections that are an order of magnitude greater than the deflection limits normally anticipated for structures under ambient conditions. Large deflections may induce forces in adjacent structural assemblies (e.g., connections).

As the strength and stiffness of a structural member is reduced under elevated temperatures, yielding or buckling modes may occur at service gravity load levels (Seif et al. 2016). For members with temperature gradients, the resulting gradient in member stiffness and strength can alter the combined axial and flexural capacity of the member.

The designer should consider the following potential fire effects when conducting an SFED:

- Changes in material stiffness and strength at elevated temperatures;
- Induced axial forces, which typically are compressive during heating and tensile during cooling and depend on restraint conditions;
- Induced flexural moments due to the restraint of thermal expansion or temperature gradients;
- Lateral displacements of structural members, such as lateral displacement of perimeter columns due to thermal expansion of heated floors;
- Local and global buckling of members at elevated temperatures due to thermally induced degradation of strength or stiffness;
- Concurrent development of thermally induced forces and changes in member resistance;
- Cross-sectional area reductions resulting from material responses, such as spalling of concrete or charring of timber sections;
- Reduced capacity of connections/fasteners at elevated temperatures; and
- Time-dependent effects such as creep due to sustained loads and elevated temperatures.

Failure modes may include (1) component failure (e.g., column buckling), (2) assembly failure (e.g., connection tear-out), (3) subsystem failure (e.g., composite floor failure), (4) partial collapse (e.g., multiple floors or columns fail and a portion of the structure remains intact), and (5) global collapse (e.g., most or all of a structure fails). More than one failure mode may occur. For example, a component or assembly failure such as column buckling or connection failure may lead to a partial or global collapse. Time-dependent evaluation of thermal effects should be performed using temperature–time histories of the structural members and assemblies derived from thermal analyses.

7.2 FIRE EFFECTS ON STRUCTURAL DEMAND AND CAPACITY

Fire effects on structural members can affect both structural demand and capacity. For example, member mechanical properties may be degraded, resulting in larger deflections or deformations that may also result in load redistributions. Thermal effects on structural systems and failure modes are presented here. Detailed information on thermally induced degradation of mechanical material properties is discussed in Chapter 8. Methods of analysis that include temperature-dependent effects on structures are described in Chapter 9.

7.2.1 Thermal Expansion and Contraction

The effects of thermal expansion (during heating) and contraction (during cooling) on the structural system should be considered by the designer. Thermal expansion and contraction of members may generate forces sufficient to cause yielding or fracture depending on the degree of restraint provided by the surrounding structural system to the thermally induced actions (Gillie et al. 2002). In fact, restrained thermal expansion may dominate the behavior of framing systems, particularly floor systems, with degradation of stiffness and strength being a secondary factor (Bailey et al. 1999).

After a fire has been extinguished, fire-exposed members that have experienced plastic deformation due to degraded mechanical properties and restraint of thermal expansion may also experience thermal contraction as the structure naturally cools. Both thermal expansion and contraction of members may induce forces in adjacent structural assemblies and connections, depending on the level of thermal restraint. In addition, thermal effects may induce additional loads in adjacent structural framing, regardless of whether the adjacent bay was directly heated by fire.

Columns typically do not experience significant thermal restraint unless they are braced or otherwise restrained from vertical elongation. Because gravity loads on columns remain essentially constant during a fire event, they do not provide any additional restraint to thermally induced elongation or contraction. Floor framing members, depending on the degree of moment-resisting connectivity, typically do not provide significant restraint of column elongation or shortening. However, columns may experience significant rotational restraint from floor framing (Agarwal et al. 2014).

Standard fire tests designate floor beams and assemblies as restrained or unrestrained. This designation refers to whether or not the thermal expansion of the test specimen is resisted by the frame of the test furnace. Actual structural systems under realistic fire exposure may experience a range of restraint conditions and magnitudes. Accordingly, designers should analyze the level of restraint from adjacent structural framing that would resist the thermal expansion of a heated assembly or subsystem and not extrapolate standard fire test results to evaluate the restraint condition of a structural system (LaMalva 2017).

7.2.2 Nonuniform Heating

Nonuniform heating of structural systems under fire exposure was also discussed in Section 6.4. Structural members that are not uniformly heated may experience a combination of thermally induced forces and moments due to restraint and thermal bowing (Dwaikat et al. 2011). A restrained member with a thermal gradient through its cross section may experience a shift of the section's center of stiffness away from its geometric center, resulting in additional bending moments (Garlock and Quiel 2007). A neutral axis shift will induce moments for axially loaded members if the member's ends are rotationally restrained. Eccentricity resulting from the thermal gradient can reduce the flexural bending capacity of beams and the flexural buckling capacity of columns (Ojeda et al. 2016, Gales et al. 2016). Columns that are rotationally unrestrained (i.e., pinned) and subjected to nonuniform heating may experience thermal bowing, which can introduce an additional secondary moment owing to increasing eccentricity of axial loading (Agarwal et al. 2014).

Analytical evaluations of individual components for fire exposure may not be capable of reproducing the effects of thermal gradients. In these cases, the designer should exercise caution and perform a conservative calculation or consider methods that are capable of representing the effects of thermal gradients explicitly (Quiel and Garlock 2010; Gales et al. 2016; Zhang et al. 2014; Agarwal et al. 2014; Choe et al. 2017). Computational structural analyses of members and assemblies under fire exposure should include thermal gradients where applicable.

7.2.3 Cooling Phase Effects

During the decay or cooling phase of a fire, additional residual forces and deformations may occur in the structural members and the connections. For example, axial tensile forces in floor beams may develop during the cooling phase of a fire as the beam contracts. Once the member cools to ambient temperature, if elastic or buckling limits have been exceeded, permanent deformation may have occurred, necessitating retrofit or replacement. The connections may also experience an increase in tensile force, potentially leading to connection damage or failure (Liu et al. 2002, Bailey et al. 1996, Hantouche et al. 2016). It should be noted that in some observed failures, partial fracture of a connection may have reduced the thermally induced lateral loading on columns and prevented column instability or avoided permanent damage to the overall structural system (Wang et al. 2012). However, such failure mechanisms should not be relied upon when conducting an SFED.

7.3 FAILURE MODES

Fire effects on a structural system may result in failure modes that differ from those that occur at ambient conditions. Depending on the level of thermal restraint and the temperature–time history, failure modes may include large or rapid (i.e., *runaway*) increases in deflection, tensile rupture, member buckling (local, global, or lateral torsional), connection failures (bearing, tear-out, bolt shear rupture, weld rupture, and so forth), reinforcement failure, or loss of section (concrete spalling, cracking or crushing, or timber charring). The following sections examine failure modes that are particular to specific materials of construction.

7.3.1 Structural Steel

AISC 360-10 Appendix 4 provides design guidance for steel structures subject to fire exposure (AISC 2016). For steel temperatures below 200°C (400°F), the degradation of steel properties due to thermal effects need not be considered. For steel temperatures greater than 200°C (400°F), the use of degraded material properties is required. Forces and deformations induced by elevated temperatures must also be considered. Gross et al. (2013) provide examples for the application of AISC 360 Appendix 4 to SFED.

Structural analyses should examine the sequence of structural responses to increasing and decreasing temperatures because the nonlinear responses may be cumulative and/or path-dependent. Simple analysis methods for determining the response of structural members and components are limited to cases where a uniform temperature can be assumed. Analysis of elements that develop thermal gradients must consider whether gradient effects on the capacity (Garlock and Quiel 2008) and mechanics (Garlock and Quiel 2007) of the element are significant for the structural design fire considered. Advanced methods may be required when considering the overall structural response to fire exposure, the interaction between structural members, or the residual strength following heating.

At elevated temperatures, the elastic modulus of steel degrades more rapidly than the yield strength. Hence, the ratio of elastic modulus to yield strength at elevated temperature artificially indicates an increase in strength. Note that the nominal member strengths provided by AISC 360 Equations E3-2 and E3-3 (AISC 2016) for ambient temperatures are not appropriate for elevated temperatures. Instead, the equations in Section 4.2.4d of AISC

Figure 7-1. Buckling of a steel column near the ceiling area.
Source: FEMA (2002).

360 Appendix 4 (AISC 2016) should be used for simplified calculations of single element capacity.

Local buckling may occur in a steel member when a portion of a cross section has a higher temperature than the rest of the member. For example, a steel column section near the ceiling of a compartment where hot gases create a *hot zone* may experience local buckling (Figure 7-1) and/or the bottom flange of a thermally restrained floor beam may have local buckling (Seif and McAllister 2013, Wang et al. 2014). Global buckling may occur if lateral restraint is lost through failure of bracing members or their connections, if there is sufficient loss of stiffness and strength over most of the member length, or if thermal gradients across the member section cause curvature sufficient to impose second-order (P-delta) forces. Lateral torsional

Figure 7-2. Lateral torsional buckling of a composite floor beam.
Source: Photo by Therese McAllister.

buckling of a composite floor beam may occur if there is significant heating of the bottom flange relative to the top flange, as shown in Figure 7-2, or if a significant number of shear stud connectors fail due to differential thermal restraint between the floor beam and concrete slab (Zhang et al. 2013). This phenomenon is not captured in standard fire testing within SFRD, in which the concrete slab and steel beam would be restrained equally during the test.

Steel connection failures may occur when component stiffness or strength is degraded at elevated temperatures (as shown in Figures 7-3 and 7-4) and/or due to axial forces or moments that develop during heating or cooling phases. During fire exposure, a concrete floor slab and its supporting steel beams may lose composite action through failure of the shear stud connections (McAllister et al. 2013). The slab and beam members would then carry the applied gravity loads separately. Large displacements or sagging may occur if the steel beam experiences significant loss of stiffness relative to the applied loads (as shown in Figure 7-5), and the steel beam may change its load-carrying mechanism from flexure to tension.

7.3.2 Reinforced Concrete and Masonry

Concrete structures may be constructed with a variety of materials, and the selection of materials may impact system performance under fire exposure. Structures may be cast-in-place, precast, or prestressed construction. They may use normal-weight, lightweight, or high-strength concrete and steel reinforcement or prestressing strands. The basic philosophy for the design of reinforced concrete and masonry structures to withstand fire effects is to provide sufficient cover concrete over the reinforcement to ensure that reinforcement temperatures remain below acceptable levels. However, the

Figure 7-3. Bolt failures in the Cardington tests.
Source: British Steel (1999).

Figure 7-4. Failed tree connections in WTC 5.
Source: FEMA (2002).

Figure 7-5. Sagging floor beams following fires.
Source: FEMA (2002).

capacity of the structural member or system to resist applied service loads during a fire event needs to be checked for expected member temperatures using temperature-dependent material properties for concrete and steel materials (see Chapter 8).

Heat-induced spalling of concrete elements may result in both a reduced cross section and loss of cover (and thus insulation protection) for steel reinforcement, as shown in Figure 7-6. The potential for concrete spalling at high temperature, as well as its potential consequences, should be carefully considered when analyzing reinforced concrete structures under fire exposure.

Normal-weight concrete is widely used; however, the choice of aggregate (carbonate or siliceous) affects its thermal performance and ability to withstand fire effects, as discussed in Section 5.5. Lightweight concrete tends to provide more thermal resistance to heating, and high-strength concrete may be prone to spalling in certain conditions, as discussed in Section 8.3.1.6.

Design guidance for precast and prestressed concrete members to withstand fire effects is similar to that for reinforced concrete (Gustaferro and Martin 1989). Notably, the designer should consider thermal protection for connections and joints, where steel components may be exposed to fire. Phan et al. (2010) provide an overview of existing design guidance for concrete members and assemblies.

Figure 7-6. Spalled concrete on a column following fire exposure. Source: Mlakar et al. (2003).

For masonry, the designer should devote special attention to cracking in grout and/or mortar between masonry units for an SFED because these areas may experience the greatest damage owing to restraint of thermal expansion. Also, the designer should consider the risk of concrete spalling, as discussed in Section 8.3.1.6.

7.3.3 Timber

Wood is a combustible material and experiences charring when exposed to fire, as discussed in Section 8.5. Structural analyses of timber members under fire exposure should account for appropriate charring rates. Although a portion of the structural cross section may be lost due to charring (as shown in Figure 7-7), thermal analysis of wood cross sections can take advantage of the insulating properties of the char layer, which can slow the temperature rise of the unaffected regions of the cross section.

Loss of structural integrity due to charring represents the primary mode of failure in wood structures (Buchanan and Abu 2017, Kukay et al. 2012). Also, the designer should devote special attention to connections, at which partial charring may result in total failure of nailed, plated, or bolted segments (many of which are notched to accommodate the interface of connected members).

7.3.4 Nonstructural Components

ASCE 7 Appendix E only applies to structural systems themselves and excludes consideration of other fire resistant assembly types including fire barriers, smoke barriers, firestop systems, and so forth (ASCE 2016). These excluded assemblies primarily serve to compartmentalize buildings during a fire event. In these cases, the mechanical integrity of the given assembly under fire exposure usually governs, as opposed to its structural response.

Figure 7-7. Fire test of glulam beams.
Source: reThink Wood (2017), photo by David Barber.

The SFPE S.02 standard defines the difference between structural response and mechanical integrity response (SFPE 2015). An example of mechanical integrity response is the delamination of gypsum wallboard from wall studs due to thermal degradation.

Because the mechanical integrity response cannot be reliably predicted using analytical tools, the evaluation of nonstructural components should only be addressed using SFRD, which is not addressed in this manual. In this case, empirical testing is deemed sufficient to properly index the fire resistance of such assemblies. Also, nonstructural hazards such as falling items (e.g., suspended ceiling and lighting) are not within the purview of this manual but may be addressed by experience and testing.

REFERENCES

Agarwal, A., L. Choe, and A. Varma. 2014. "Fire design of steel columns: Effects of thermal gradients." *J. Constr. Steel Res.* 93, 107–118.

AISC. 2016. *Specification for structural steel buildings*. ANSI/AISC 360-10. Chicago: AISC.

ASCE/SEI (Structural Engineering Institute). 2016. *Minimum design loads and associated criteria for buildings and other structures*. ASCE/SEI 7. Reston, VA: ASCE.

Bailey, C., W. Burgess, and R. Plank. 1996. "Analyses of the effects of cooling and fire spread on steel-framed buildings." *Fire Saf. J.* 26(4), 273–293.

Bailey, C., T. Lennon, and D. Moore. 1999. "The behaviour of full-scale steel-framed buildings subjected to compartment fires." *Struct. Eng.* 77(8), 15–21.

British Steel. 1999. "The behaviour of multi-storey steel framed buildings in fire." *A European Joint Research Program*. South Yorkshire, UK: Swinden Technology Centre.

Buchanan, A., and A. Abu. 2017. *Structural design for fire safety*, 2nd Ed. New York: Wiley.

CEN. 2005. *Eurocode 3—Design of steel structures. Part 1–2: General rules—Structural fire design*. EN 1993-1-2. Brussels: CEN.

Choe, L., C. Zhang, W. E. Luecke, and A. H. Varma. 2017. "Influence of material models on predicting the fire behavior of steel columns." *Fire Technol.* 53(1), 375–400.

Dwaikat, M., V. Kodur, S. Quiel, and M. Garlock. 2011. "Experimental behavior of steel beam–columns subjected to fire-induced thermal gradients." *J. Constr. Steel Res.* 67(1), 30–38.

FEMA. 2002. *World Trade Center building performance study: Data collection, preliminary observations, and recommendations*. Washington, DC: FEMA.

Gales, J., K. Hartin, and L. Bisby. 2016. *Structural fire performance of contemporary post-tensioned concrete construction*. New York: Springer. DOI 10.1007/978-1-4939-3280-1_2.

Garlock, M., and S. Quiel. 2007. "Mechanics of wide-flanged steel sections that develop thermal gradients due to fire exposure." *Steel Struct.* 7, 153–162.

Garlock M., and S. Quiel. 2008. "Plastic axial load and moment interaction curves for fire-exposed steel sections with thermal gradients." *J. Struct. Eng.* 134(6), 874–880.

Gillie, M., A. Usmani, and J. Rotter. 2002. "A structural analysis of the Cardington British Steel corner test." *J. Constr. Steel Res.* 58(4), 427–442.

Gross, J., N. Iwankiw, and M. Johan. 2013. "Structural fire engineering: Overview and application examples of Appendix 4 of the AISC specification." *AISC Eng. J.* Q4, 273–289.

Gustaferro, A., and L. Martin. 1989. *Design for fire resistance of precast prestressed concrete*. Chicago: Prestressed Concrete Institute.

Hantouche, E., N. Abboud, M. Morovat, and M. Engelhardt. 2016. "Analysis of steel bolted double angle connections at elevated temperatures." *Fire Saf. J.* 83, 79–89.

Kukay, B., R. White, and F. Woeste. 2012. "Inspection, testing, and monitoring of buildings and bridges." In *Fire damage of wood structures*. Washington, DC: International Code Council.

LaMalva, K. 2017. "Reexamination of 'restrained vs. unrestrained'." *FPE Extra*, 20. Bethesda, MD: Society of Fire Protection Engineers.

Liu, T., M. Fahad, and J. Davies. 2002. "Experimental Investigation of Behavior of Axially Restrained Steel Beams in Fire." *J. Constr. Steel Res.* 58(9), 1211–1230.

McAllister, T., J. Gross, F. Sadek, S. Kirkpatrick, R. MacNeill, M. Zarghamee, et al. 2013. "Structural response of World Trade Center buildings 1, 2 and 7 to impact and fire damage." *Fire Technol.* 49(3), 709–739.

Mlakar, P., D. Dusenberry, J. Harris, G. Haynes, L. Phan, and M. Sozen. 2003. "Pentagon building performance report." Reston, VA: ASCE.

Ojeda, O., J. Maljaars, and R. Abspoel. 2016. "Fire-exposed steel columns with a thermal gradient over the cross-section." *Thin-Walled Struct.* 98, 103–110.

Phan, L., T. McAllister, J. Gross, and M. Hurley. 2010. *NIST TN 1681: Best practice guidelines for structural fire resistance design of concrete and steel buildings*. Gaithersburg, MD: NIST.

Quiel, S., and M. Garlock. 2010. "Closed-form prediction of the thermal and structural response of a perimeter column in a fire." *Open Construct. Build. Technol. J.* 4(1), 64–78.

reThink Wood. 2017. "Glulam connections fire test report now available." *Cision PR Newswire*. Accessed June 26, 2018. https://www.prnewswire

.com/news-releases/glulam-connections-fire-test-report-now-available-300507431.html.

Seif, M., and T. McAllister. 2013. "Stability of wide flange structural steel columns at elevated temperatures." *J. Constr. Steel Res.* 84, 17–26.

Seif, M., J. Main, J. Weigand, F. Sadek, L. Choe, C. Zhang, et al. 2016. *Temperature-dependent material modeling for structural steels: Formulation and application*. Gaithersburg, MD: NIST.

SFPE (Society of Fire Protection Engineers). 2015. *Engineering standard on calculation methods to predict the thermal performance of structural and fire resistive assemblies*. SFPE S.02. Gaithersburg, MD: Society of Fire Protection Engineers.

Wang, Y., I. Burgess, F. Wald, and M. Gillie. 2012. *Performance-based fire engineering of structures*. Boca Raton, FL: CRC Press.

Wang, W., V. Kodur, X. Yang, and G. Li. 2014. "Experimental study on local buckling of axially compressed steel stub columns at elevated temperatures." *Thin-Walled Struct.* 82, 33–45.

Zeng, D., K. Lo, K. Cheang, and J. Lai. 2013. "Remaining strengths and pitting resistance of AISI 316 after a fire attack: Implications for use as concrete rebars." JMEPEG 22, ASTM International (https://slideheaven.com/remaining-strengths-and-pitting-resistance-of-aisi-316-after-a-fire-attack-impli.html).

Zhang, C., J. Gross, and T. McAllister. 2013. "Lateral torsional buckling of steel W-beams subjected to localized fires." *J. Constr. Steel Res.* 88, 330–338.

Zhang, C., L. Choe, and Z. Zhang. 2014. "Behavior of axially loaded steel short columns subjected to a localized fire." *J. Constr. Steel Res.* 111, 103–111.

CHAPTER 8
MECHANICAL MATERIAL PROPERTIES

8.1 OVERVIEW

This chapter provides an overview of how the mechanical properties (e.g., elastic modulus, yield strength, and ultimate strength) of steel, concrete, masonry, and wood are affected by temperature. Some fundamental concepts about the effects of temperature on the mechanical properties of materials are covered, and references to the most widely accepted material models found in design codes and the research literature are provided. Available information is provided for completeness, and in some cases a range of experimental data from the literature is presented to show the degree of uncertainty in material properties at elevated temperatures. For many practical applications, the designer may default to one of the material models found in codes and standards (e.g., Eurocode, AISC, ACI); these models were designed to provide a suitable level of conservatism and have been thoroughly vetted within their intended scope of application.

It is recommended that the designer be judicious in the selection of material models for structural fire engineering design (SFED), keeping in mind the application that is under consideration:

- For forensic-level applications, which seek to match empirical observations, for instance, the designer is advised to use mechanical properties that match the stress–strain–temperature response of the actual materials used in construction as closely as possible. This may involve, for example, the use of *actual strength* rather than *nominal strength* in the stress–strain–temperature model.
- For design purposes, the designer is advised to use mechanical properties that provide some degree of conservatism. For instance, the

lower-bound estimates of strength and Young's modulus may be most appropriate. These may be determined from the contents of this chapter and/or from the research literature.

SFED analyses may be very sensitive to the material properties that are assumed. Therefore, it is good practice to perform a sensitivity analysis before selecting design values that are deemed critical to the SFED. Note that the response of materials to elevated temperatures is an active area of research, and in some cases (e.g., predicting spalling in concrete), there is a lack of industry consensus.

To model the temperature-dependent mechanical response of structural systems at elevated temperatures, the designer, at a minimum, should define a temperature-dependent stress–strain response for the material. This generally requires the specification of both (a) elastic response and (b) plastic behavior beyond the yield point as a function of temperature. Any material model referenced will require the values of both Young's modulus and Poisson's ratio to describe the linear, elastic behavior. Because the values of Poisson's ratio for different structural materials are relatively insensitive to changes in temperature, it is recommended that the designer use ambient values for SFED (CEN 2006, AISC 2016).

The nonlinear plastic behavior (i.e., beyond the yield point) can be described by providing yield stress as a function of strain at elevated temperatures. Predicting large deformations and fracture of elements using finite element programs requires information on the true stress–strain response of the material up to very large strains. A true stress–strain curve can be determined from an experimentally measured engineering stress–strain curve in a straightforward manner before the onset of necking for ductile materials (e.g., steel). However, after the onset of necking, determining the true stress–strain curve is more difficult because of the nonuniform deformation and complex state of stress and strain within the necked region of a tension test coupon.

Fire exposure can have residual effects on the stress–strain response of materials following cooling. The term *residual strength* herein refers to the strength of the material after it is heated by fire exposure to a specified temperature and then cooled to ambient temperature. SFED may involve analysis of the structural system following a fire event (e.g., analysis of structural elements that support building refuge areas). In these cases, the designer should define material properties based on the temperature that was reached during the fire event and, in some cases, the rate at which it was cooled (e.g., cooled by convection with air versus in water). Sections below provide residual material properties accordingly.

The designer should be aware that the data reported in the literature are often influenced by the mechanical testing regime used (e.g., steady-state versus transient-state test) and the assumptions made in defining the yield strength of the material as a function of strain. Hence, the careless use of temperature-dependent stress–strain data could result in significant errors in an SFED. When modeling steel, for example, it is important for the designer to carefully consider how the yield strength is defined (e.g., using 0.2% offset strain or 2% total strain) and whether creep effects are implicitly included in the model or not. Similarly, the stress–strain–temperature response of concrete and masonry is highly dependent on the aggregate type, and the behavior of wood depends on its species, grade, and other contributing factors (e.g., direction of the loading). Accordingly, the designer should be familiar with the relevant factors that affect the response of materials at elevated temperatures to avoid significant modeling errors.

8.2 MECHANICAL PROPERTIES OF STEEL

8.2.1 Structural Steel

Important mechanical properties of structural steel such as yield stress, ultimate strength, proportional limit, and modulus of elasticity are all established using the fundamental stress–strain behavior at elevated temperatures. A variety of tests can be conducted to characterize the stress–strain behavior of steel at elevated temperatures, but the tension test is the most common.

8.2.1.1 Stress–Strain–Temperature Response. Under fire exposure, the stress–strain behavior of structural steel becomes time-dependent, especially for temperatures at or above 500°C (Morovat et al. 2012, Morovat 2014). For SFED applications, it is common practice to account for time effects indirectly through either loading rates or temperature rates. It is also possible to explicitly quantify time effects on the stress–strain behavior of structural steel when exposed to elevated temperatures.

8.2.1.1.1 Rate- and Temperature-Dependent Response. As mentioned previously, time effects on the stress–strain behavior of structural steel at elevated temperatures are often quantified indirectly as rate effects. To evaluate rate effects, two general types of material tests are performed. The first type of test is referred to as the steady-state temperature tension test, in which under constant temperature conditions, the time effects are implicitly considered through loading rates. The second type of test is referred to as the transient-state temperature tension test, in which under constant load conditions, time effects are implicitly considered through temperature rates. Choe et al. (2017) compared the mechanical properties of structural steels developed by these two different types of tension tests and evaluated the effects of stress–strain models used in the finite element models of structural steel columns at elevated temperatures.

8.2.1.1.1.1 Steady-State Temperature Tension Test. In steady-state temperature tension tests, the steel coupon temperature is first increased to the test temperature under no load (i.e., no restraint to thermal expansion). Static load is then monotonically increased while the temperature is held constant. The rate at which the static load is applied to the coupon is therefore the main factor affecting the mechanical behavior of steel obtained in a steady-state temperature tension test. For a specified temperature, steady-state temperature tension tests directly result in stress–strain curves corresponding to different loading rates. Note that most reported steel properties at elevated temperatures are based on steady-state temperature tension testing because of its simplicity and practicality (Outinen et al. 2001, Outinen and Mäkeläinen 2007, Lee et al. 2003, Lee et al. 2013). Luecke et al. (2011) provide a compilation of steady-state temperature tensions tests on structural steels.

Representative stress–strain curves determined from steady-state temperature tension tests of ASTM A992 structural steel specimens at various temperatures are shown in Figure 8-1. All the stress–strain curves shown in Figure 8-1 are from tests conducted at a constant crosshead displacement rate of 0.01 in./min. Figure 8-2 further plots the initial parts of the stress–strain curves in Figure 8-1 up to 0.5% strain. As shown in Figures 8-1 and 8-2, the fundamental shape of the stress–strain curve changes as temperature increases. Specifically, at 400°C and above, structural steel no longer exhibits a well-defined yield plateau, and it demonstrates significant nonlinearity at low levels of stress and strain. This

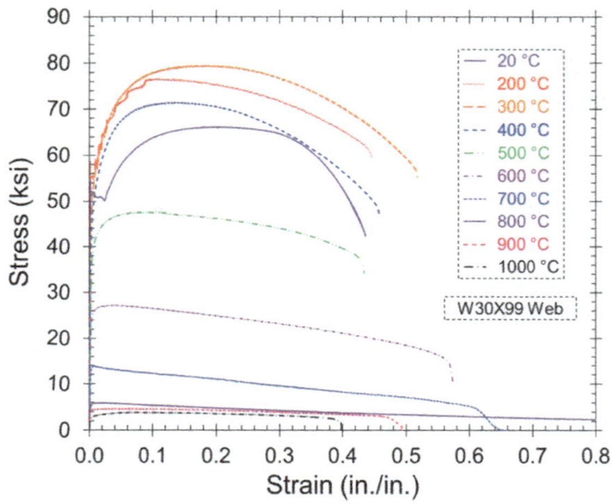

Figure 8-1. Engineering stress–strain curves of ASTM A992 steel at elevated temperatures obtained from steady-state temperature tension tests (entire curves). Source: Lee et al. (2013).

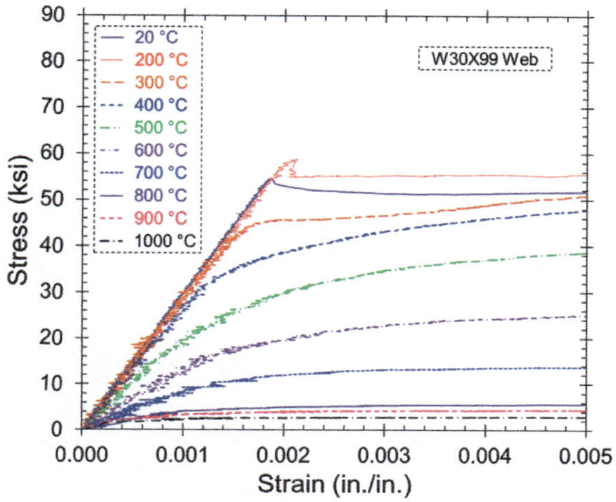

Figure 8-2. Engineering stress–strain curves of ASTM A992 steel at elevated temperatures obtained from steady-state temperature tension tests (initial portion of curves).
Source: Lee et al. (2013).

early nonlinearity may be particularly significant when considering the stability of a structure under fire exposure, wherein tangent stiffness is a critical material property. In addition, Figures 8-1 and 8-2 clearly show that the yield stress and modulus of elasticity decrease with the increase in temperature.

As stated, the loading rate is the major factor affecting the stress–strain curves of structural steel that are derived from a steady-state temperature tension test. For example, Figure 8-3 represents the effect of loading rate on the stress–strain curves of ASTM A992 structural steel determined from the steady-state temperature tension tests. The results are specifically shown for two constant crosshead displacement rates of 0.01 in./min and 0.10 in./min and two different temperatures of 600°C and 700°C. To illustrate the loading rate effects clearly, only the initial portions of the stress–strain curves up to 2% strain are plotted in Figure 8-3. The impact of the lower loading rate in reducing the yield and tensile strengths of ASTM A992 steel at elevated temperatures is clearly evident in Figure 8-3. More specifically, a displacement rate of 0.01 in./min results in yield and tensile strengths that are both 30% to 40% lower than those obtained at 0.10 in./min for the considered temperatures. Also, the lower loading rate results in higher nonlinearity in the *knee region* of the stress–strain curve of structural steel. This higher nonlinearity particularly affects both the buckling capacity and the thermally induced forces in steel columns subjected to elevated temperatures. Seif et al. (2016a) and Luecke et al. (2011) discuss a temperature-dependent stress–strain model that accounts for strain rate sensitivity. Overall, the

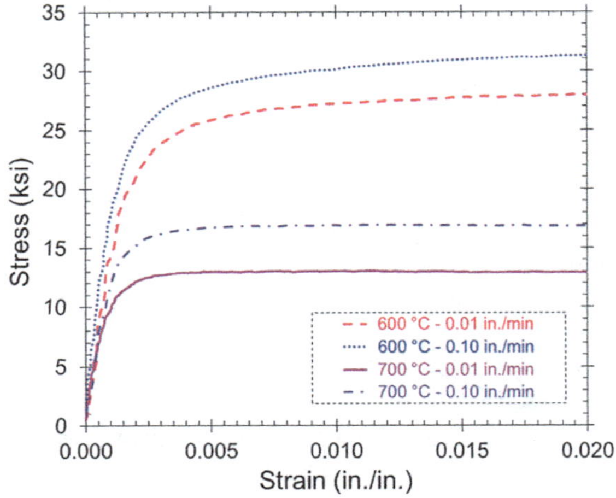

Figure 8-3. Loading rate effect on the engineering stress–strain curves of ASTM A992 steel at elevated temperatures obtained from steady-state temperature tension tests.
Source: Lee et al. (2013).

significant change in mechanical properties owing to variable loading rates is an indication of the time-dependent behavior of structural steel at elevated temperatures.

8.2.1.1.1.2 Transient-State Temperature Tension Test. In transient-state temperature tension tests, the steel coupon is first loaded to a target engineering stress level at ambient temperature. While holding the stress constant, the temperature is increased until fracture of the coupon occurs. Hence, the rate at which the temperature is increased is the main factor affecting the mechanical behavior of the steel. For a specified engineering stress, transient-state temperature tension tests result in strain–temperature curves corresponding to different heating rates. As a result, transient-state temperature tension tests do not directly result in stress–strain curves. A procedure that involves subtracting thermal strains from total measured strains and applying the cross-plotting technique is used to convert the strain–temperature curves into stress–strain curves at elevated temperatures (as shown in Figure 8-4). Details on the conversion can be found in the literature (Kirby and Preston 1988; Outinen et al. 2001; Outinen and Mäkeläinen 2007; Chen et al. 2006; Schneider and Lange 2010).

Representative strain–temperature curves for Grade 50B structural steel derived from transient-state temperature tension tests at various applied stresses are shown in Figure 8-5. All strain–temperature curves presented in Figure 8-4 are from tests conducted at a constant temperature rate of 10°C/min.

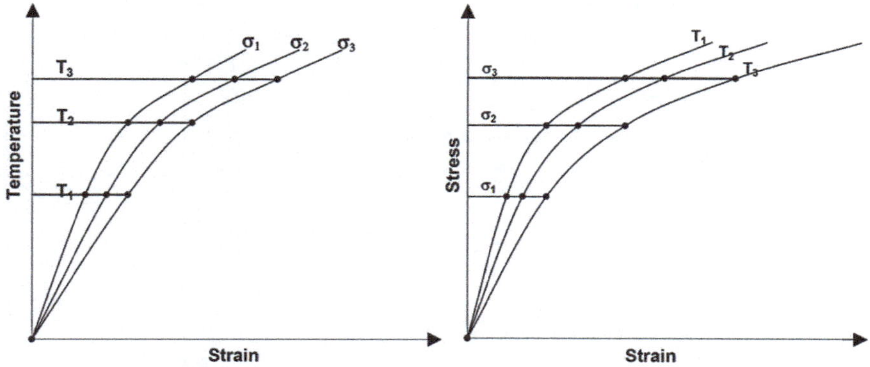

Figure 8-4. Constructing engineering stress–strain curves using strain–temperature curves obtained from transient-state temperature tension tests.
Source: Outinen and Mäkeläinen (2007).

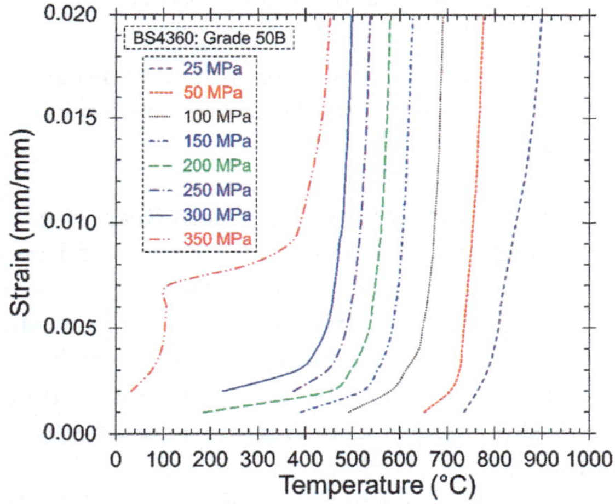

Figure 8-5. Engineering strain–temperature curves of Grade 50B steel obtained from transient-state temperature tension tests.
Source: Kirby and Preston (1988).

As stated, the heating rate can affect the strain–temperature curves of structural steel measured in a transient-state temperature tension test. For example, Figure 8-6 illustrates the effect of heating rate on the strain–temperature curves of Grade 50B structural steel. Specifically, the results are shown for two constant temperature rates of 10°C/min and 20°C/min, and two different stress levels of 150 MPa and 350 MPa. For these tests, the impact of the higher heating rate in reducing the strains reached at a

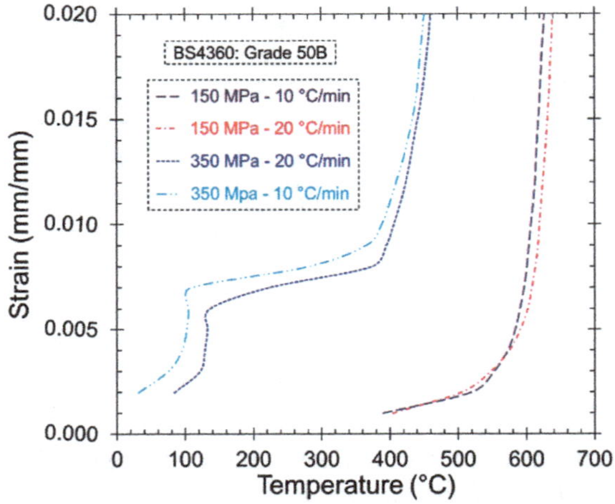

Figure 8-6. Temperature rate effect on engineering strain–temperature curves of Grade 50B steel obtained from transient-state temperature tension tests. Source: British Steel (1983), Kirby and Preston (1988).

certain temperature is modest. However, time-dependent effects on the stress–strain behavior of steel should be accounted for when a transient-state temperature tension test is used.

8.2.1.1.2 Time- and Temperature-Dependent Response. In the previous section, material characterization tests were introduced to implicitly quantify the effect of time on the elevated-temperature behavior of steel for use in SFED. If required, the time effects on the stress–strain behavior of structural steel at elevated temperatures can be also directly evaluated. To explicitly account for time effects, two primary tension tests are conducted: the steady-state temperature creep test and the steady-state temperature relaxation test.

8.2.1.1.2.1 Steady-State Temperature Creep Test. The most common material characterization test to quantify the time-dependent behavior of structural steel at elevated temperatures is the steady-state temperature creep test in tension. This test is a force-controlled test in which the steel coupon temperature is first increased to the test temperature under no load. A load that produces a specific engineering stress is then quickly applied while the temperature is held constant. Both the stress and the temperature are subsequently held constant, and the strain is monitored for a specific period of time. At a specific temperature, steady-state temperature creep tests result in strain–time curves corresponding to different engineering stresses. A typical creep strain versus time curve is shown in Figure 8-7 (Morovat 2014). This curve is often divided into the three phases of primary,

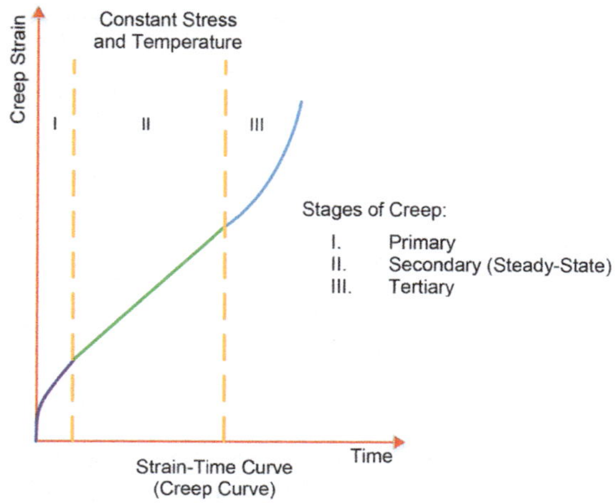

Figure 8-7. Qualitative representation of structural steel creep.
Source: Morovat (2014).

secondary, and tertiary creep. In the primary stage, the curve is nonlinear and typically exhibits a decreasing creep strain rate with increasing time. In the secondary stage, the creep strain rate is constant, often referred to as steady-state creep. In the tertiary stage, the creep strain rate increases with time. For structural steel, the shape of the strain–time curve and the magnitude of the creep strain (time-dependent strain) are highly dependent on temperature and stress level.

There are a limited number of studies on the thermal creep of structural steel with direct application to SFED (Harmathy 1967, 1970; Knight, Skinner, and Lay 1971; Skinner 1972; Fujimoto et al. 1980; Fujimoto et al. 1981; Fujimoto et al. 1982; Luecke 2005). More recently, experimental creep data on ASTM A992 and ASTM A572 structural steels were collected for temperatures, times, and stress levels of interest in SFED (Morovat et al. 2012, Lee 2012, Lee et al. 2012a, Kodur and Aziz 2015). Figure 8-8 depicts sample results of creep tests on samples of ASTM A992 steel. Specifically, the measured creep strain versus time response of the steel after being exposed to constant stresses at 700°C is shown. Creep effects are present at temperatures on the order of 700°C and stress levels of 4.5 ksi. However, the curve corresponding to the stress of 8.7 ksi has much larger creep strains in the same time frame of 1 h.

8.2.1.1.2.2 Steady-State Temperature Relaxation Test. Steady-state temperature relaxation testing in tension is another material characterization test used to quantify the time-dependent behavior of structural steel at elevated temperatures. This test is a strain-controlled (displacement-controlled) test in which the steel coupon temperature is first increased to the test

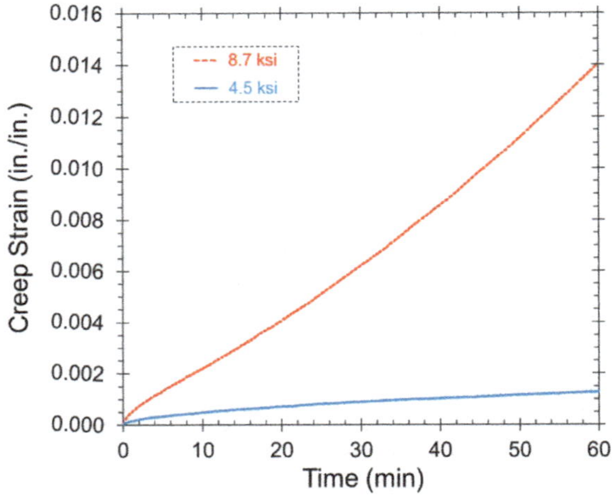

Figure 8-8. Representative results of creep tests on samples of ASTM A992 steel at 700°C.
Source: Morovat (2014).

temperature under no load. The steel coupon is then strained rapidly to a specific engineering strain level while temperature is held constant. Both the strain and the temperature are subsequently held constant and the stress is monitored for a specific time period. At a specific temperature, steady-state temperature relaxation tests result in stress–time curves corresponding to different engineering strain levels.

An experimental study conducted on the elevated-temperature stress relaxation of steel examined with temperature levels (up to 1,000°C) and durations (minutes up to several hours) of interest in SFED (Lee 2012). A typical relaxation curve measured in this study is shown in Figure 8-9. At a temperature of 500°C, the steel exhibited a large relaxation of stress (i.e., from about 44 ksi to 30 ksi), indicating a significant influence of time on the material response of structural steel at elevated temperatures.

8.2.1.1.3. Stress–Strain Models in Design Codes (Design Models). As discussed in previous sections, the stress–strain behavior of structural steel at elevated temperatures is significantly different from that at ambient temperature. Specifically, the fundamental shape of the stress–strain curve becomes highly nonlinear at the *knee region*, mainly owing to the disappearance of a distinct yield plateau at high temperatures that reduces the proportional limit. Accordingly, several design codes provide equations that represent the stress–strain curves for structural steel at elevated temperatures. For instance, the equations provided by Eurocode 3 (CEN 2006) for the elevated-temperature stress–strain behavior for structural steel along

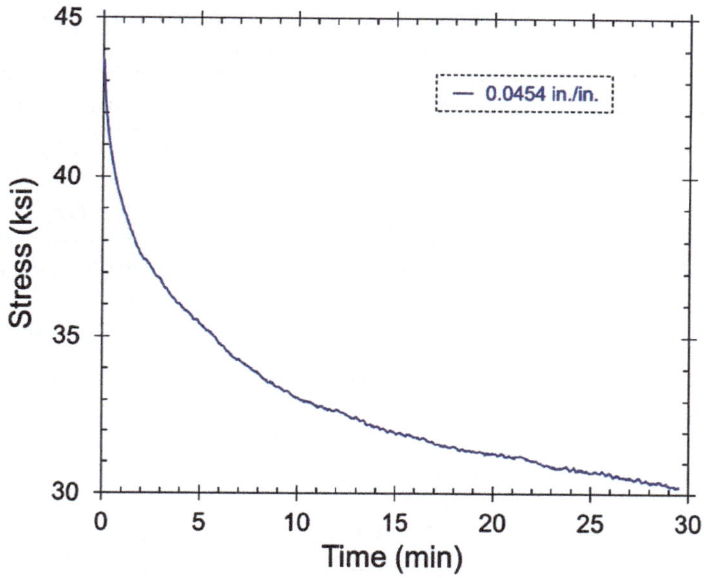

Figure 8-9. A representative relaxation curve for ASTM A992 steel at 500°C. Source: Lee (2013).

with the defining parameters are summarized in Table 8-1. Note that AISC 360 (2016) does not provide stress–strain models for structural steel at high temperatures. Further, as observed from the stress–strain equations listed in Table 8-1, Eurocode 3 (CEN 2006) does not consider the effects of thermal creep explicitly but includes contributions implicitly.

8.2.1.2 Mechanical Properties at Elevated Temperatures. A common method adopted in design codes to evaluate the mechanical properties of structural steel at elevated temperatures is to apply reduction factors to such properties at ambient temperature. For example, Eurocode 3 (CEN 2006) provides reduction factors for important mechanical properties of structural steel at elevated temperatures such as yield strength, proportional limit, and elastic modulus. These temperature-dependent reduction factors are listed in Table 8-2. It should be noted that in Eurocode 3 (CEN 2006), the yield strength for structural steel at elevated temperatures is defined as the stress corresponding to the 2% total strain and as a result, this stress is termed the *effective yield strength*.

At ambient temperature, the proportional limit of steel is assumed to be equal to the yield strength. At elevated temperatures, the proportional limit occurs at a stress that is less than the yield strength. This reflects the fact that the stress–strain curves for structural steel at elevated temperatures become highly nonlinear at stresses well below the yield stress. Note that

Table 8-1. Stress–Strain Relationships for Structural Steel at Elevated Temperatures.

Strain Range	Stress σ	Tangent Modulus
$\varepsilon \le \varepsilon_{p,T}$	εE_T	E_T
$\varepsilon_{p,T} < \varepsilon < \varepsilon_{y,T}$	$F_{p,T} - c + (b/a)\sqrt{a^2 - (\varepsilon_{y,T} - \varepsilon)^2}$	$\dfrac{b(\varepsilon_{y,T} - \varepsilon)}{a\sqrt{a^2 - (\varepsilon_{y,T} - \varepsilon)^2}}$
$\varepsilon_{y,T} \le \varepsilon \le \varepsilon_{t,T}$	$F_{y,T}$	0
$\varepsilon_{t,T} < \varepsilon < \varepsilon_{u,T}$	$F_{y,T}\left[1 - \dfrac{(\varepsilon - \varepsilon_{t,T})}{(\varepsilon_{u,T} - \varepsilon_{t,T})}\right]$	—
$\varepsilon = \varepsilon_{u,T}$	0	—
Parameters	$\varepsilon_{p,T} = F_{p,T}/E_T$; $\varepsilon_{y,T} = 0.02$; $\varepsilon_{t,T} = 0.15$; $\varepsilon_{u,T} = 0.02$	
	$a^2 = (\varepsilon_{y,T} - \varepsilon_{p,T})(\varepsilon_{y,T} - \varepsilon_{p,T} + c/E_T)$	
Functions	$b^2 = c(\varepsilon_{y,T} - \varepsilon_{p,T})\, E_T + c^2$	
	$c = \dfrac{(F_{y,T} - F_{p,T})^2}{(\varepsilon_{y,T} - \varepsilon_{p,T})\, E_T - 2(F_{y,T} - F_{p,T})}$	

where
E_T = elastic modulus at temperature T
$F_{y,T}$ = effective yield strength at temperature T
$F_{p,T}$ = proportional limit at temperature T
$\varepsilon_{y,T}$ = strain at yield at temperature T
$\varepsilon_{p,T}$ = strain at proportional limit at temperature T
$\varepsilon_{t,T}$ = limit strain for yield strength at temperature T
$\varepsilon_{u,T}$ = ultimate strain at temperature T

Source: CEN (2006).

in Table 8-2, there are no reductions in yield strength until the temperature exceeds 400°C. However, the elastic modulus and proportional limit are reduced for temperatures in excess of only 100°C.

For predicting the mechanical properties of structural steel at elevated temperatures, AISC 360 also adopts the method of applying reduction factors to the corresponding ambient-temperature mechanical properties

Table 8-2. Reduction Factors for Mechanical Properties of Structural Steel at Elevated Temperatures Per Eurocode 3.

	Reduction Factor at Temperature (T) to the Values of f_y and E at 20°C		
Steel Temperature (°C)	Reduction Factor for Yield Strength $k_{y,T}=F_{y,T}/F_y$	Reduction Factor for Proportional Limit $k_{p,T}=F_{p,T}/F_y$	Reduction Factor for Elastic Modulus $k_{E,T}=E_T/E$
20	1.0000	1.0000	1.0000
100	1.0000	1.0000	1.0000
200	1.0000	0.8070	0.9000
300	1.0000	0.6130	0.8000
400	1.0000	0.4200	0.7000
500	0.7800	0.3600	0.6000
600	0.4700	0.1800	0.3100
700	0.2300	0.0750	0.1300
800	0.1100	0.0500	0.0900
900	0.0600	0.0375	0.0675
1,000	0.0400	0.0250	0.0450
1,100	0.0200	0.0125	0.0225
1,200	0.0000	0.0000	0.0000

Source: CEN (2006).

(AISC 2016). These reduction factors are listed in Table 8-3. Aside from small differences owing to rounding, AISC 360 essentially has the same reduction factors as Eurocode 3.

As seen in both Table 8-2 and Table 8-3, the key mechanical properties of structural steel including yield strength, proportional limit, and modulus of elasticity all decrease as the temperature increases. Also, the reduction rates of strength and elastic modulus vary significantly in different temperature ranges. At temperatures lower than 400°C, the reduction is relatively small, whereas at higher temperatures, the material degradation is much more significant.

More recently, a temperature-dependent stress–strain model for structural steels was formulated based on steady-state test data for 42 individual structural steels, augmented by the post-yield behavior derived from eight of those steels (Seif et al. 2016a). The model accounts for the change in yield strength and elastic modulus with temperature, the change in the amount of post-yield strain hardening with both temperature and ambient-temperature yield strength, and the change in strength with increasing strain rate. The model is extensible to other steels, such as fire-resistive structural steels

Table 8-3. Reduction Factors for Mechanical Properties of Structural Steel at Elevated Temperatures Per AISC Specification 360.

Steel Temperature (°C/°F)	Reduction Factor at Temperature (T) to the Values of f_y and E at 20°C (65°F)		
	Reduction Factor for Yield Strength $k_{y,T}=f_{y,T}/f_y$	Reduction Factor for Proportional Limit $k_{p,T}=f_{p,T}/f_y$	Reduction Factor for Elastic Modulus $k_{E,T}=E_T/E$
20/68	1.00	1.00	1.00
93/200	1.00	1.00	1.00
200/400	1.00	0.80	0.90
320/600	1.00	0.58	0.78
400/750	1.00	0.42	0.70
430/800	0.94	0.40	0.67
540/1,000	0.66	0.29	0.49
650/1,200	0.35	0.13	0.22
760/1,400	0.16	0.06	0.11
870/1,600	0.07	0.04	0.07
980/1,800	0.04	0.03	0.05
1,100/2,000	0.02	0.01	0.02
1,200/2,200	0.00	0.00	0.00

Source: AISC (2016).

and quenched-and-tempered plate steels, either by substituting a new description for the retained yield strength or by fitting new parameters for the entire stress–strain behavior.

8.2.1.3 Residual Properties. Following an uncontrolled fire event, there may be a need to evaluate the remaining load-carrying capacity of structural steel members to establish the safety of the structural system and to assess the need for repair or replacement of damaged elements (Tide 1998, Gosain et al. 2008). Also, the structural stability of occupant refuge areas may need to be evaluated (see Section 3.1). Accordingly, knowledge of the material properties of structural steel after exposure to elevated temperatures and subsequent cooling is fundamental. In particular, the rate of cooling is one important factor known to affect the mechanical behavior of heat-treated steels. The actual cooling rates for structural steel after a fire exposure depend on a number of factors, including the cooling rate itself, whether or not the steel is insulated, and whether or not the steel is exposed to water from fire-fighting operations or sprinklers.

There are limited experimental data that pertain to post-fire evaluation of structural steel (Smith et al. 1981, Outinen and Mäkeläinen 2007, Lee et al. 2012b, Lee 2012). For example, representative stress–strain curves obtained from tension tests on samples of ASTM A992 steel that were subjected to heating and cooling are shown in Figure 8-10 through Figure 8-13 (Lee 2012, Lee et al. 2012b). The results are shown for two specific cooling methods: cooled-in-air (CIA) (Figures 8-10 and 8-11) and cooled-in-water (CIW) (Figures 8-12 and 8-13). The data in Figures 8-10 and 8-11 indicate that for temperatures up to about 500°C, the stress–strain curves for the heated and cooled samples are very similar to that of the unheated steel. However, for heating temperatures of 600°C and above, some changes from the virgin steel become somewhat more noticeable.

For the heating range of 600°C to 1,000°C, the CIA-resulted stress–strain data show some reduction in yield and tensile strength. However, the stress–strain curves maintain the same basic shape of the virgin curve, exhibiting a well-defined yield plateau. Within the 600°C to 1,000°C heating range, more significant changes are evident in the CIW-resulted stress–strain data, as shown in Figures 8-12 and 8-13. Compared to those of the virgin steel, the CIW-resulted stress–strain curves exhibit a significant increase in tensile strength and a significant loss of ductility. For heating temperatures in the range of 800°C to 1,000°C, the CIW-resulted stress–strain curves do not exhibit a yield plateau and show significant nonlinearity starting at relatively low stress levels. In general, more significant changes in the stress–strain curves are expected for steel heated in the range of 800°C to 1,000°C because structural steel undergoes a phase change near 730°C from ferrite (α-Fe) to austenite (γ-Fe).

8.2.2 High-Strength Bolts

The behavior of structural bolts at elevated temperatures affects the performance of beam-end steel framing connections under fire exposure. Structural bolts are made of high-strength steel, which contains different alloys and is subjected to different heat treatments as compared to structural steel. As a result, structural bolts behave differently than structural steel when exposed to elevated temperatures. Accordingly, designers should use information contained in this section in lieu of that presented in Section 8.2.1 for structural steel bolts.

The strength of structural bolts at elevated temperatures can be determined from bolt shear tests and bolt tension tests. Also, tension tests can be performed on coupons machined from high-strength bolts to obtain stress–strain curves, which can be used to derive the tensile strength of bolts at elevated temperatures. Hardness tests may also be conducted on bolts to provide an estimate of tensile strength. In addition, hardness tests can

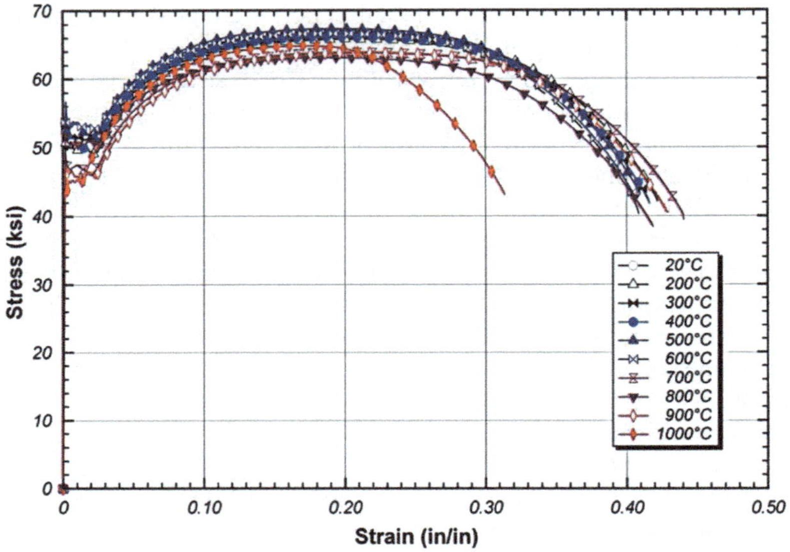

Figure 8-10. CIA-resulted engineering stress–strain curves for ASTM A992 steel subjected to different elevated temperatures (entire curves).
Source: Lee (2012).

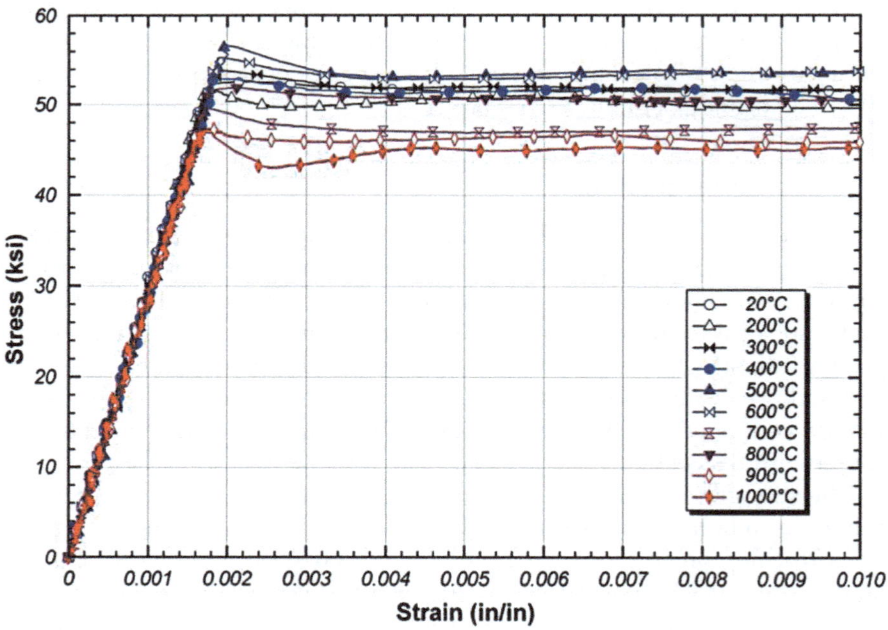

Figure 8-11. CIA-resulted engineering stress–strain curves for ASTM A992 steel subjected to different elevated temperatures (initial portion of curves).
Source: Lee (2012).

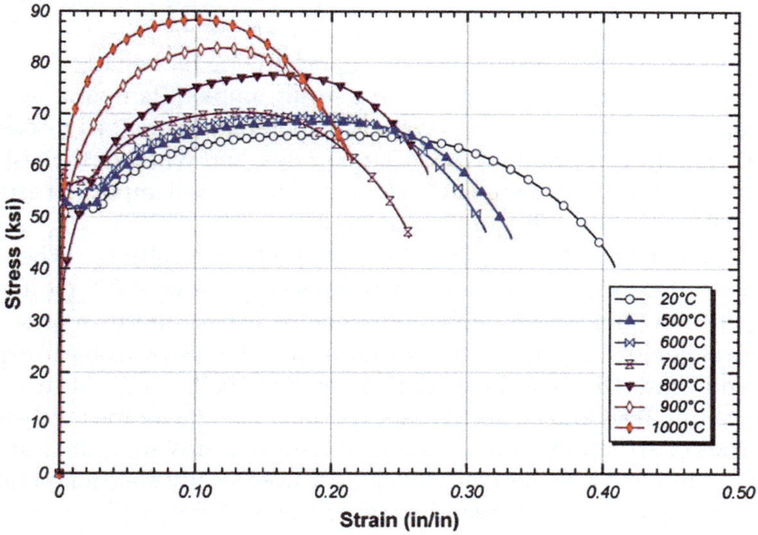

Figure 8-12. CIW-resulted engineering stress–strain curves for ASTM A992 steel subjected to different elevated temperatures (entire curves).
Source: Lee (2012).

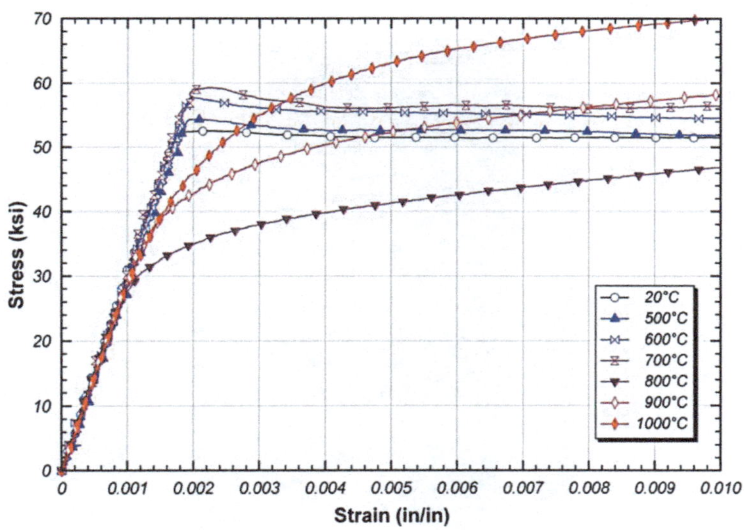

Figure 8-13. CIW-resulted engineering stress-strain curves for ASTM A992 steel subjected to different elevated temperatures (initial portion of curves).
Source: Lee (2012).

provide an economical and convenient way to assess the residual strength of bolts after a fire.

There is no standard test method for characterizing the behavior of structural bolts at elevated temperatures. As a result, ambient-temperature tests are adapted and modified for elevated-temperature testing, as discussed herein. Therefore, heating methods, strain rates, and whether the test fixtures are protected from the elevated temperatures are all important aspects in evaluating the reliability of the test results.

The strength of high-strength bolts at elevated temperatures can become time-dependent, especially for temperatures at or above 400°C (Seif et al. 2016b). In SFED, time effects on the bolt strength can be indirectly represented through either loading rates or temperature rates. It is also possible to explicitly quantify time effects on the strength of structural bolts exposed to elevated temperatures. It should be noted that experimental data for the mechanical properties of structural bolts at elevated temperatures are quite limited. Furthermore, almost all the available data are from steady-state temperature tests, in which time effects are indirectly accounted for using loading rates.

8.2.2.1 Shear Strength of Bolts. The shear strength of structural bolts at elevated temperatures can be evaluated by conducting either single-shear tests or double-shear tests (Figure 8-14). These tests have been mostly performed under steady-state temperature conditions (Kirby 1995; Yu 2006, 2009; Hanus et al. 2011; Kand 2011; Kodur et al. 2012). After the bolt specimen is first heated to the desired temperature, that temperature is held constant and the load is

Figure 8-14. Elevated-temperature double-shear tests of high-strength bolts. Source: Yu (2006).

applied monotonically until failure. The rate at which the static load is applied to the bolt specimen is the primary factor affecting the mechanical behavior of structural bolt obtained in a steady-state temperature shear test.

Figures 8-15 and 8-16 show sample results of elevated-temperature double-shear tests conducted on samples of A325 and A490 bolts, respectively. The results of double-shear tests are specifically presented in the form of load–displacement curves at different temperatures. Note that the shear capacity of the bolts is half of the maximum load recorded in double-shear tests at elevated temperatures. For both the A325 and A490 high-strength bolts, the stiffness, strength, and shape of the load–displacement curves in shear all change with temperature. More specifically, Figure 8-15, clearly indicates that the shear load capacity of A325 bolts drops significantly at 400°C. Figure 8-15 further shows the ductile behavior of A325 bolts in shear for temperatures at and above 500°C. Similar trends in the shear behavior of A490 bolts at elevated temperatures can be seen in Figure 8-16. It is important to note that load–displacement curves in Figure 8-15 show a 0.2-in. displacement at the beginning of the tests, indicating slip prior to the application of the actual shear load, and further initial displacement at elevated temperatures owing to thermal elongation and loss of bolt pretension. As a result of slip, the load–displacement ratio for 25°C appears to be less than that for 400°C. Therefore, the designer should remove the slip offsets from the measured displacements

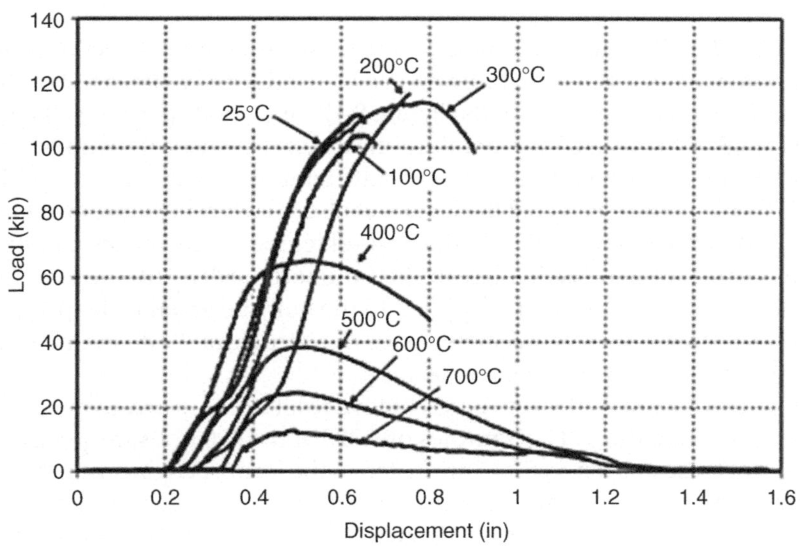

Figure 8-15. Load-displacement curves for ASTM A325 bolts obtained in double shear tests at elevated temperatures.
Source: Yu (2006, 2009).

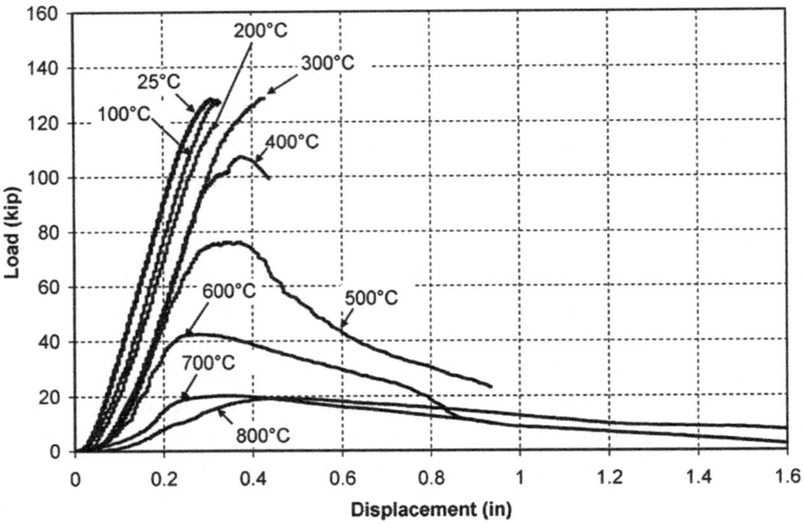

Figure 8-16. Load-displacement curves for ASTM A490 bolts obtained in double shear tests at elevated temperatures.
Source: Yu (2006, 2009).

to develop stress–strain curves or failure criteria for structural bolts as part of an SFED.

8.2.2.2 Tensile Strength of Bolts. The tensile strength of structural bolts at elevated temperatures can be evaluated by conducting tension tests either on machined coupons from bolts (Figure 8-17) or on bolt specimens (Figure 8-18). Similar to elevated-temperature shear tests, tensile tests of bolts have been mostly performed under steady-state temperature conditions (Kirby 1995; Gonzalez and Lange 2009; Hanus et al. 2011; Kand 2011; Kodur et al. 2012). After the bolt is heated to the desired temperature, the temperature is held constant and the load is applied monotonically until failure. The rate at which the static load is applied to the bolt specimen is the primary factor affecting the mechanical behavior of a structural bolt obtained in a steady-state temperature tension test.

Figures 8-19 and 8-20 show representative results of elevated-temperature tension tests conducted on samples of A325 and A490 bolts, respectively. Results of these tension tests are presented in the form of engineering stress–strain curves at different temperatures. Note that the maximum stress recorded in these tension tests indicates the tensile capacity of the bolts at elevated temperatures. For both A325 and A490 high-strength bolts, the stiffness, strength, and shape of the stress–strain curves in tension all change with temperature. More specifically, Figures 8-19 and 8-20 both indicate that

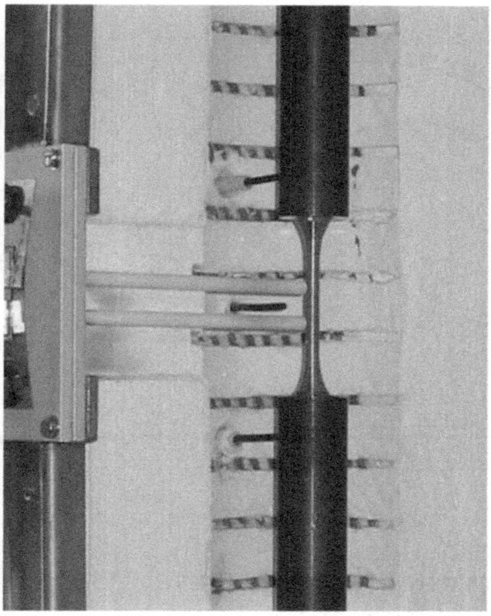

Figure 8-17. Elevated-temperature tension tests on machined coupon from high-strength bolts.
Source: Gonzalez and Lange (2009).

Figure 8-18. Elevated-temperature tension test on high-strength bolt specimen.
Source: Gonzalez and Lange (2009).

the tensile strength of A325 and A490 bolts decreases significantly above 400°C and that the strain corresponding to the ultimate stress decreases with temperature. Comparing the engineering stress–strain curves shown in Figure 8-1 for structural steel and those presented in Figures 8-19 and 8-20 for structural bolts further indicates that high-strength bolts, being quenched

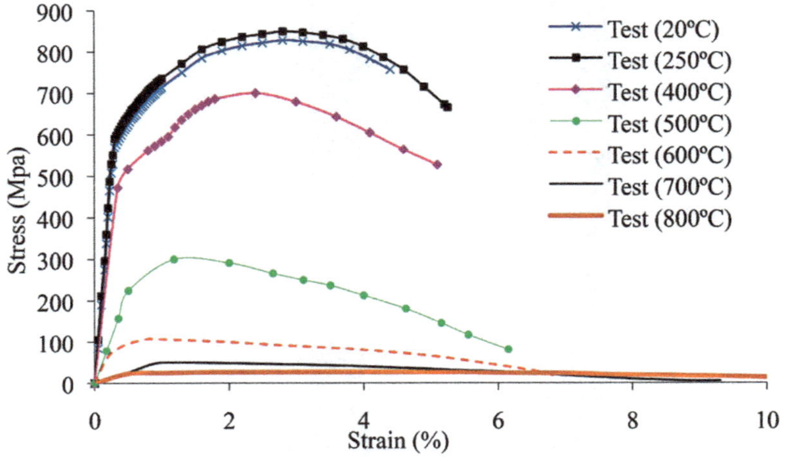

Figure 8-19. Engineering stress–strain curves for ASTM A325 bolts obtained in tension tests at elevated temperatures.
Source: Kodur et al. (2012).

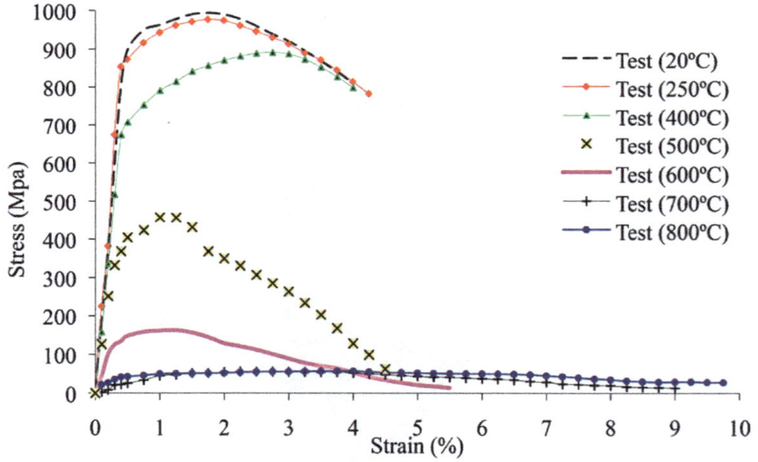

Figure 8-20. Engineering stress–strain curves for ASTM A490 bolts obtained in tension tests at elevated temperatures.
Source: Kodur et al. (2012).

and tempered, lose stiffness and strength more rapidly than structural steel at temperatures above 400°C. This has been demonstrated in other studies as well (e.g., Seif et al. 2016b).

As discussed previously, time effects on the tensile strength of structural bolts at elevated temperatures can be directly evaluated if required. For

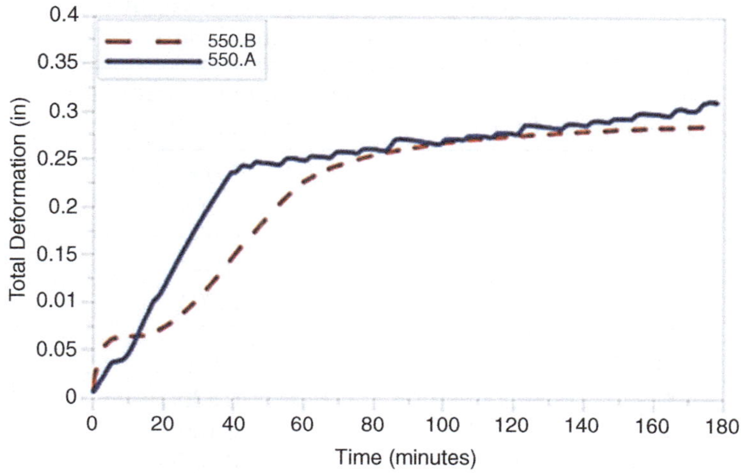

Figure 8-21. Time-dependent behavior of ASTM A325 bolts obtained in tension tests with a constant load of 2 kips at 550°C.
Source: Matar (2014).

instance, the steady-state temperature creep test in tension can be used. In this test, the bolt temperature is first increased to the test temperature under no load. A load is then applied very fast while the temperature is held constant. Both the load and the temperature are consequently held constant, and the displacement is monitored for specific period of time. At a specific temperature, steady-state temperature creep tests result in displacement–time curves corresponding to different applied loads.

There are only a limited number of studies on the thermal creep of structural bolts with application to SFED. Figure 8-21 depicts sample results of creep tests on ASTM A325 bolts (Matar 2014). These tests were conducted under the applied load of 2 kips at 550°C. As seen in Figure 8-21, the elongation (deformation) of ASTM A325 bolts can be significant at elevated temperatures.

8.2.2.3 Bolt Strength in Design Codes. A common method adopted in design codes to evaluate the strength of structural bolts at elevated temperatures is to apply reduction factors to ambient-temperature strengths. For example, Annex D of Eurocode 3 (CEN 2006) provides strength reduction factors for structural bolts at elevated temperatures up to 1,000°C, as listed in Table 8-4. Eurocode 3 (CEN 2006) recommends that the same reduction factors be used for both tension and shear loading of bolts. From Table 8-4, it can be observed that bolt strength is assumed to decrease at temperatures of 100°C and above, and it decreases to zero when 1,000°C is reached.

Table 8-4. Reduction Factors for Bolt Strength at Elevated Temperatures per Eurocode 3.

Temperature (°C)	Reduction factor for bolt strength ($k_{b,T}$)
20	1.000
100	0.968
150	0.952
200	0.935
300	0.903
400	0.775
500	0.550
600	0.220
700	0.100
800	0.067
900	0.033
1,000	0.000

Source: CEN (2006).

In addition to Eurocode 3 (CEN 2006), AISC 360 (AISC 2016) proposes reduction factors to adjust the strength of structural bolts at elevated temperatures. The reduction factors provided in AISC 360 are listed in Table 8-5. Aside from small differences due to rounding, AISC 360 and Eurocode 3 provide essentially the same reduction factors for the strength of bolts at elevated temperatures.

Comparing the reduction factors provided in Eurocode 3 (CEN 2006) and AISC 360 (AISC 2016) for structural steel (Table 8-2 and Table 8-3) and those provided for structural bolts (Table 8-4 and Table 8-5) indicates that bolts suffer greater loss of strength at elevated temperatures compared to structural steel. For example, the Eurocode 3 (CEN 2006) strength reduction factors for structural steel at 500°C, 600°C, and 700°C are 0.78, 0.47, and 0.23, respectively, and the corresponding strength reduction factors for bolts are 0.55, 0.22, and 0.10, respectively.

8.2.2.4 Residual Properties. Following an uncontrolled fire event, there may be a need to evaluate the remaining load-carrying capacity of steel-bolted connections to establish the safety of the structural system and to assess the need for repair or replacement of damaged elements (Tide 1998, Gosain et al. 2008). Also, the structural stability of occupant refuge areas may need to be evaluated (see Section 3.1). Accordingly, knowledge of the material properties of structural bolts after exposure to elevated temperatures and subsequent cooling is fundamental.

Table 8-5. Reduction Factors for Bolt Strength at Elevated Temperatures per AISC 360.

Temperature (°C/°F)	Reduction factor for bolt strength $F_{nt}(T)/F_{nt}$ or $F_{nv}(T)/F_{nv}$
20/68	1.00
93/200	0.97
150/300	0.95
200/400	0.93
320/600	0.88
430/800	0.71
480/900	0.59
540/1,000	0.42
650/1,200	0.16
760/1,400	0.08
870/1,600	0.04
980/1,800	0.01
1,100/2,000	0.00

Source: AISC (2010).

The residual strength of structural bolts following exposure to elevated temperatures can be evaluated by different test methods at ambient temperature including bolt shear tests, bolt tension tests, and hardness tests. However, there are very limited experimental data on the post-fire mechanical properties of high-strength structural bolts (Kirby 1995; Yu 2006; Yu and Frank 2009; Lou et al. 2010; Hanus 2011). For instance, Yu (2006) investigated the residual shear capacities of A325 and A490 bolts after exposure to temperatures up to 800°C by conducting direct shear tests and hardness tests after heating and cooling of the bolts (Figure 8-22). Representative data for the residual shear capacities predicted through ambient-temperature shear tests are shown in Figures 8-23 and 8-24 for both types of bolts. When heated above about 400°C and subsequently cooled to ambient temperature, it can be seen that the high-strength bolts suffered a permanent loss of strength.

Yu (2006) also conducted a study to verify the effect of cooling rate on bolt residual strength. Samples of A325 bolts were heated to and maintained at 600°C for about 4 h. They were then cooled in cold water (for about 5 s), in air (for 30 min), and inside the furnace (for more than 14 h). The results indicated that the cooling rate had very limited impact on the residual tensile strength of ASTM A325 bolts (Yu 2006, Yu and Frank 2009).

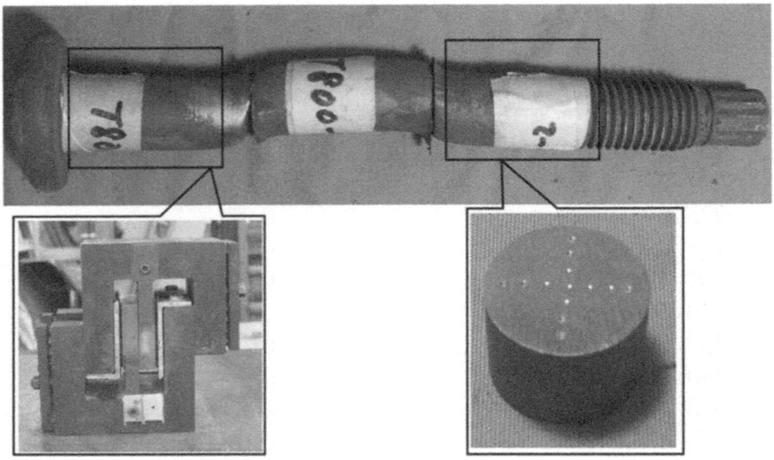

Figure 8-22. Direct shear test and hardness test for post-fire evaluations of bolt strength.
Source: Yu (2006, 2009).

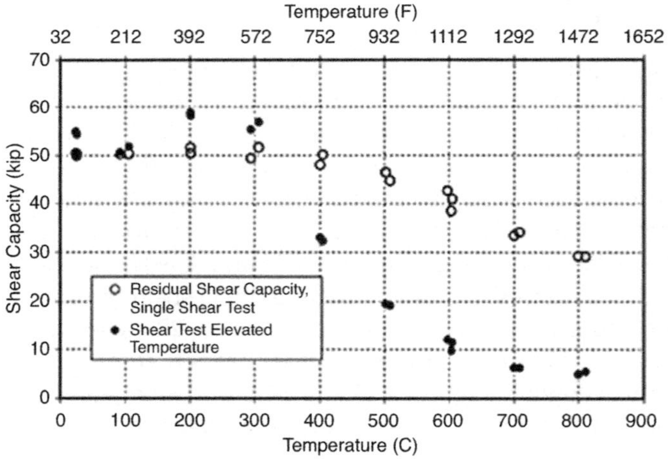

Figure 8-23. Shear strength of ASTM A325 bolts at elevated temperatures and after cooling down.
Source: Yu (2006, 2009).

8.2.3 High-Strength Steel

Steels with yield strengths in the range of 70 ksi to 100 ksi are classified as high-strength steel, whereas steels with yield strengths greater than 100 ksi are referred to as very-high- or ultra-high-strength steel. High-strength steels are produced by incorporating either special chemical

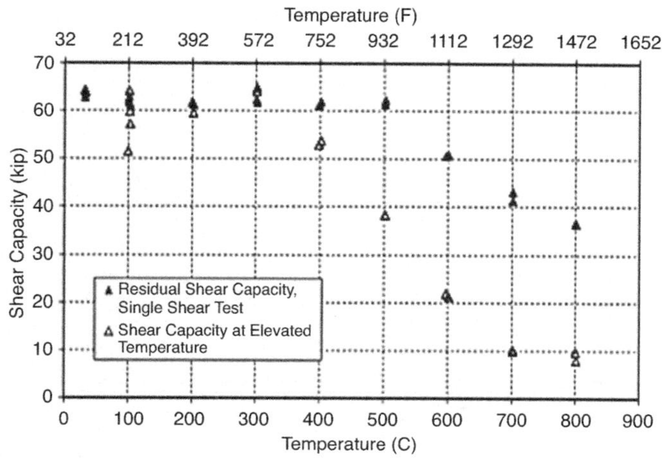

Figure 8-24. Shear strength of ASTM A490 bolts at elevated temperatures and after cooling down.
Source: Yu (2006, 2009).

compositions and/or heat treatments during rolling processes, including quenching and self-tempering. In the case of special rolling processes, the initial rapid heating and quenching in a cooling bath results in steels with higher strength. Then the following reheating and slow cooling helps the high-strength steel retain ductility. Therefore, like high-strength bolts, high-strength steels are likely to have a greater reduction in strength when heated than typical as-rolled or mild steels. Accordingly, the designer should use information contained in this section when analyzing high-strength steel in lieu of that presented in Section 8.2.1 for structural steel.

8.2.3.1 Stress–Strain–Temperature Response. Similar to structural steel, it is common practice to account for the time effects on the behavior of high-strength steel indirectly through either loading rates or temperature rates. It is also possible to explicitly quantify time effects on the stress–strain behavior of high-strength structural steel at elevated temperatures.

8.2.3.1.1. Rate- and Temperature-Dependent Response. Section 8.2.1.1.1 provides a discussion on the evaluation of rate effects of structural steel, which is applicable to high-strength steel.

8.2.3.1.1.1 Steady-State Temperature Tension Test. Section 8.2.1.1.1.1 provides a discussion on steady-state temperature tension testing of structural steel, which is applicable to high-strength steel.

Representative stress–strain curves determined from steady-state temperature tension tests on Q690 very-high-strength steel specimens at various temperatures are shown in Figure 8-25. All these stress–strain curves are

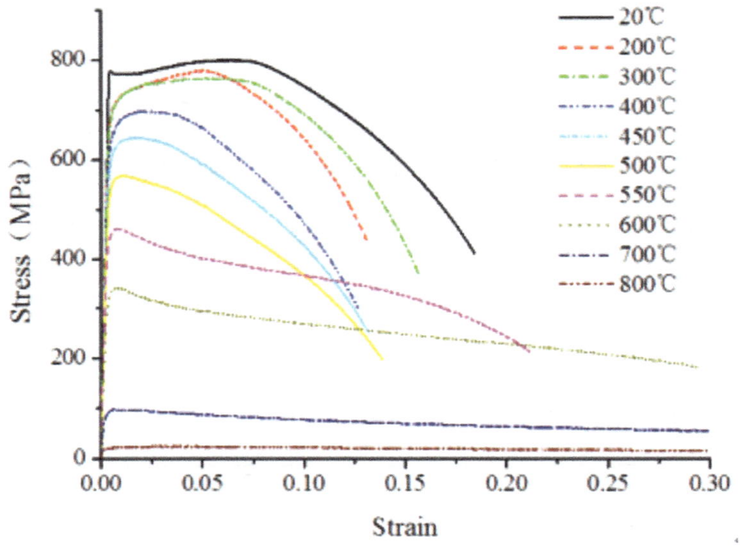

Figure 8-25. Engineering stress–strain curves of Q690 very-high-strength steel at elevated temperatures obtained from steady-state temperature tension tests (entire curves).
Source: Huang (2017).

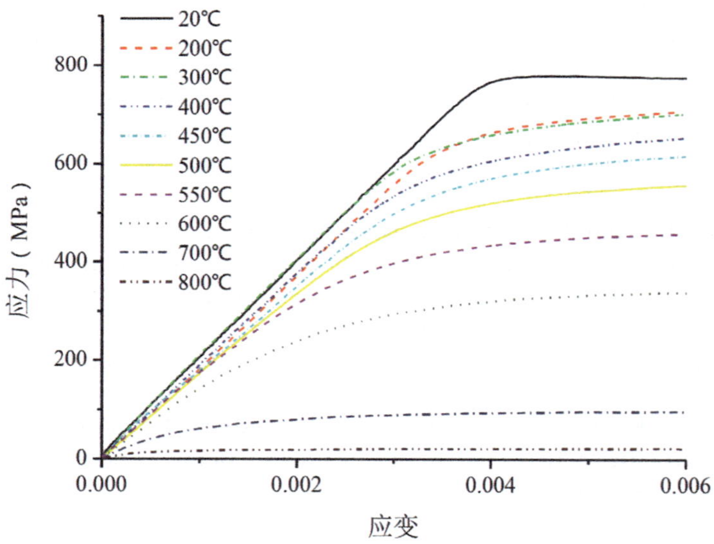

Figure 8-26. Engineering stress–strain curves of Q690 very-high-strength steel at elevated temperatures obtained from steady-state temperature tension tests (initial portion of curves).
Source: Huang (2017).

from tests conducted at a constant strain rate of 0.003/min. Figure 8-26 further plots the initial parts of the stress–strain curves in Figure 8-25 up to 0.6% strain. As shown in Figures 8-25 and 8-26, the fundamental shape of the stress–strain curve changes as temperature increases. Specifically, at 200°C and above (compared to 400°C for normal-strength steels), high-strength steel no longer exhibits a well-defined yield plateau, and it demonstrates significant nonlinearity at low levels of stress and strain. This early nonlinearity may be particularly significant when considering the stability of a structure under fire exposure, wherein tangent stiffness is a critical material property. In addition, Figures 8-25 and 8-26 both show that the yield stress and modulus of elasticity decrease with the increase in temperature.

8.2.3.1.1.2 Transient-State Temperature Tension Test. Section 8.2.1.1.1.2 provides a discussion on transient-state temperature tension testing of structural steel, which is applicable to high-strength steel. Representative engineering stress–strain curves for S960 very-high-strength structural steel derived from transient-state temperature tension tests are shown in Figure 8-27 for different temperatures. The stress–strain curves presented in Figure 8-27 are obtained from transient-state temperature tests conducted at a constant temperature rate of 10°C/min. As seen in Figure 8-27, to obtain smooth stress–strain curves at elevated temperatures, transient-state temperature tests under many different constant stresses should be conducted.

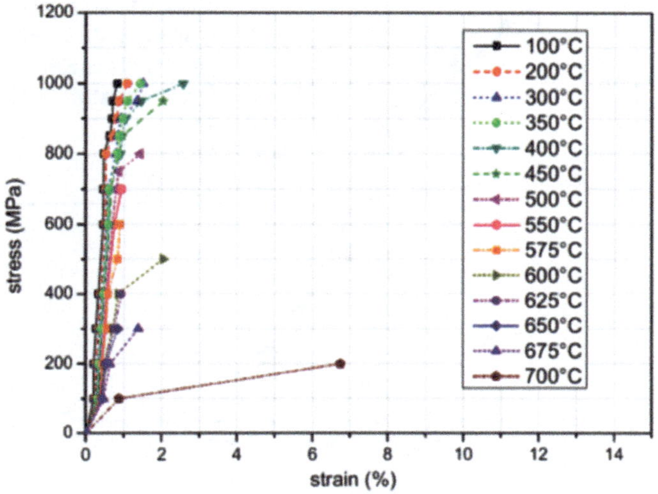

Figure 8-27. Engineering stress–strain curves of S960 very-high-strength steel obtained from transient-state temperature tension tests.
Source: Qiang et al. (2016).

8.2.3.2 Mechanical Properties at Elevated Temperatures. Section 8.2.1.2 provides a discussion on reduction factors of structural steel per design codes. It should be noted that the reduction factors proposed by Eurocode 3 (CEN 2006) and other codes are based on test results on mild- or low-carbon steel at elevated temperatures. Therefore, the designer should exercise caution when applying code-based reduction factors to high-strength steels.

To verify the accuracy of code-based reduction factors in predicting the mechanical properties of high-strength steels at elevated temperatures, Figure 8-28 and Figure 8-29, compare the code-based reduction factors with the corresponding factors from tests on different types of high-strength steel. As seen in these figures, as a result of differences in chemical composition, microstructure, and manufacturing procedure, there is scatter in the reduction factors among various high-strength steel grades. This scatter is more significant in the yield strength data. In addition, Figure 8-29 clearly shows that the code-based reduction factors overestimate the yield strength of high-strength steels at temperatures below 500°C. More important, the results shown in Figures 8-28 and 8-29 reinforce the idea that the designer should exercise caution when applying code-based reduction factors to high-strength steels.

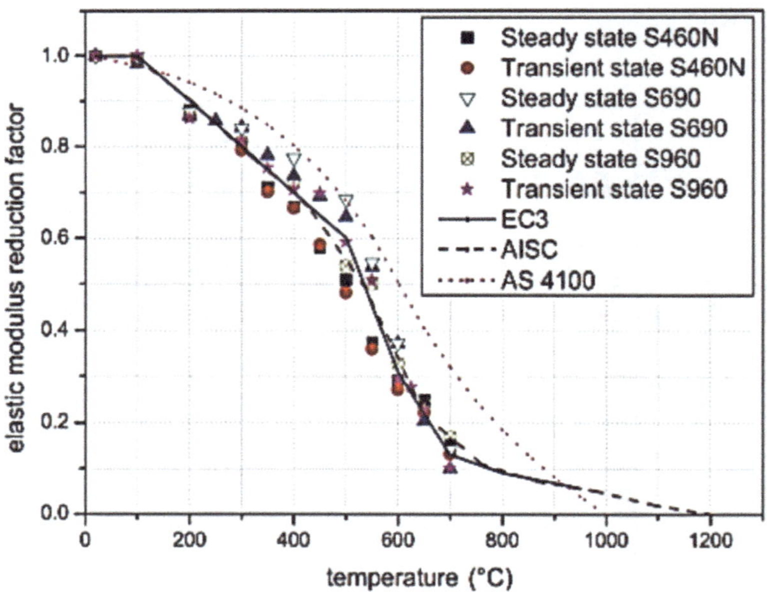

Figure 8-28. Reduction factors for elastic modulus of S460N, S690, and S960 steels at elevated temperatures.
Source: Qiang (2016).

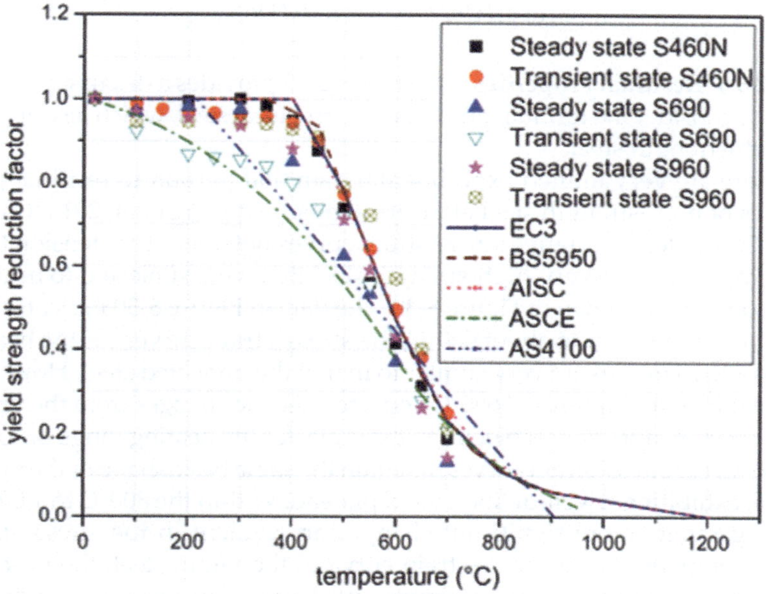

Figure 8-29. Reduction factors for yield strength of S460N, S690, and S960 steels at elevated temperatures at 2.0% total strain.
Source: Qiang (2016).

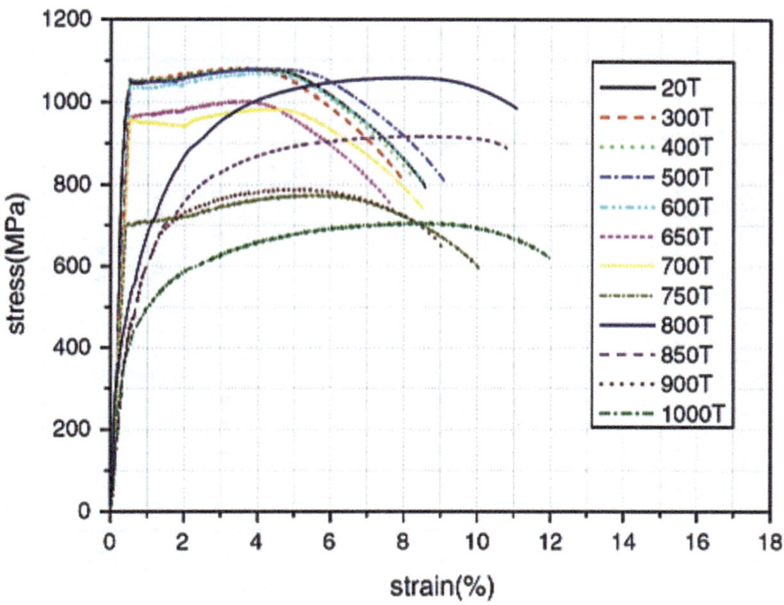

Figure 8-30. Engineering stress–strain curves for S960 very-high-strength steel following exposure to different elevated temperatures.
Source: Qiang et al. (2013).

8.2.3.3 Residual Properties. Section 8.2.1.3 provides a discussion on the need to consider the residual properties of structural steel, which also applies to high-strength steel.

There are very limited experimental data that pertain to residual properties of high-strength steel after fire exposure (Qiang et al. 2013, Chiew et al. 2014). Representative stress–strain curves obtained from tension tests on samples of S960 very-high-strength steel that were subjected to heating and cooling are shown in Figure 8-30. The data in Figure 8-30 indicate that, for temperatures up to about 600°C, the stress–strain curves for the heated and cooled samples are very similar to that of the unheated steel. However, for heating temperatures of 650°C and above, some changes from the virgin steel become more noticeable. As an example, for the heating range of 300°C to 750°C, the stress–strain curves maintain the same basic shape of the virgin curve, exhibiting a well-defined yield plateau. Within the 800°C to 1,000°C heating range, more significant changes are evident in the stress–strain data. Compared to the stress–strain curve of the virgin steel, the post-fire stress–strain curves exhibit a significant decrease in tensile strength. In addition, for heating temperatures in the range of 800°C to 1,000°C, the post-fire stress–strain curves do not exhibit a yield plateau and show significant nonlinearity starting at relatively low stress levels. As mentioned previously, more significant changes in the stress–strain curves are expected for steel

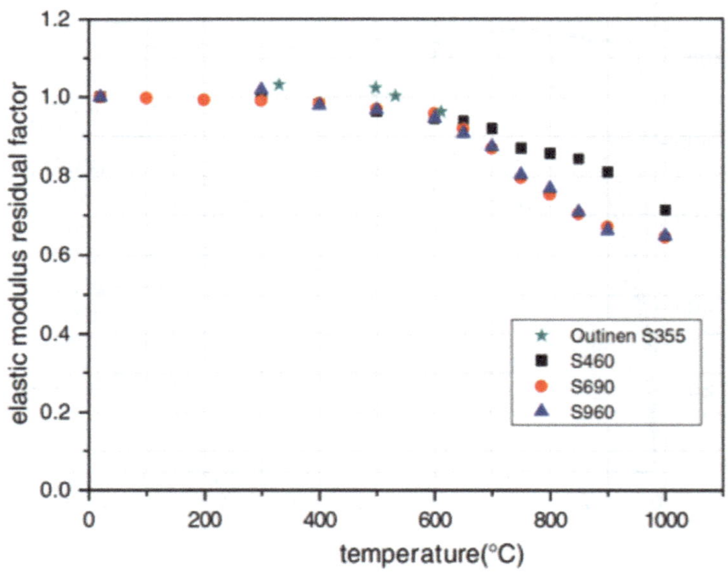

Figure 8-31. Residual elastic modulus of S460N, S690, and S960 steels following exposure to different elevated temperatures.
Source: Qiang et al. (2013).

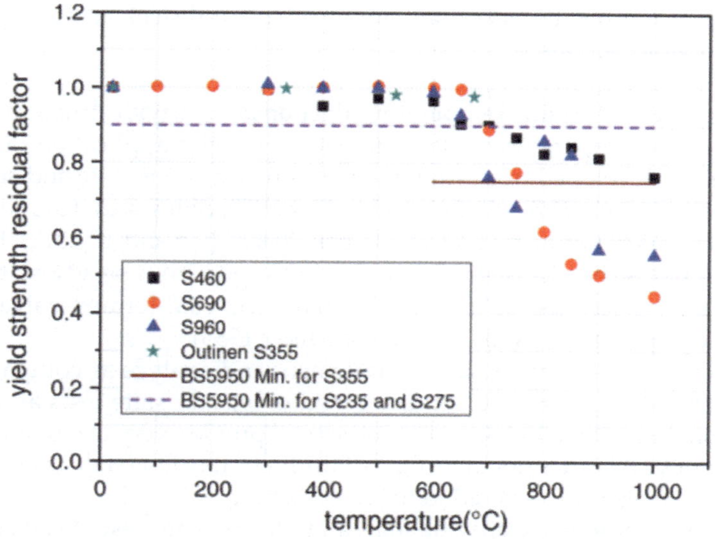

Figure 8-32. Residual yield strength of S460N, S690, and S960 steels following exposure to different elevated temperatures at 2.0% total strain.
Source: Qiang et al. (2013).

heated in the range of 800°C to 1,000°C because steel undergoes a phase change near 730°C from ferrite (α-Fe) to austenite (γ-Fe).

Figures 8-31 and 8-32 further examine the effect of preheating on the mechanical properties of high-strength steels at ambient temperature. As seen in these figures, there are significant reductions in the mechanical properties of high-strength steels when they are exposed to temperatures at or above 700°C.

8.2.4 Cold-Formed Steel

Cold-formed steel (CFS) products are produced from steel sheets or strips, usually not exceeding 3.2 mm in thickness. Common shapes in load-bearing applications are C- and Z-shapes, usually with stiffened lips, thin-walled tubular shapes, floors, and roof decks. In North America, the AISI S100 (2016) standard regulates the design of CFS structural members at ambient temperatures and defines the applicable steel grades.

In the last four decades, a substantial number of experimental and analytical studies have been conducted that have explored the tensile properties of CFS exposed to elevated temperatures. Although the chemical composition of steel grades used for CFS products is very similar to that of structural steel grades, CFS in many experiments repeatedly demonstrated a more pronounced reduction in strength (lower-strength retention factors) at elevated temperatures as compared to structural steel, as discussed in the following sections.

8.2.4.1 Stress–Strain–Temperature Response. As with structural steel, stress–strain–temperature relationships for CFS with application to SFED are usually developed from studies combining transient-state and steady-state test data sets (Mäkeläinen and Miller 1983; Sidey and Teague 1988; Zhao et al. 2005; Outinen and Mäkeläinen 2007; Chen and Ye 2012). Figure 8-33 compares the stress–strain curves developed from steady-state and transient-state tests on 0.7-mm-thick Grade Z32 cold-formed galvanized sheet steel specimens (Mäkeläinen and Miller 1983).

Steady-state tests are easier to implement and analyze as compared to transient-state tests. Hence, studies based on steady-state tests alone are more common. Nonetheless, such studies still provide very useful information about CFS grades not investigated through transient-state tests. These steady-state tests are covered in the next section.

Studies combining transient-state and steady-state tests (Mäkeläinen and Miller 1983, Sidey and Teague 1988) led to the incorporation of a CFS stress–strain–temperature model into the European Convention for Constructional Steelwork (ECCS) Model Code on Fire Engineering (ECCS 2001).

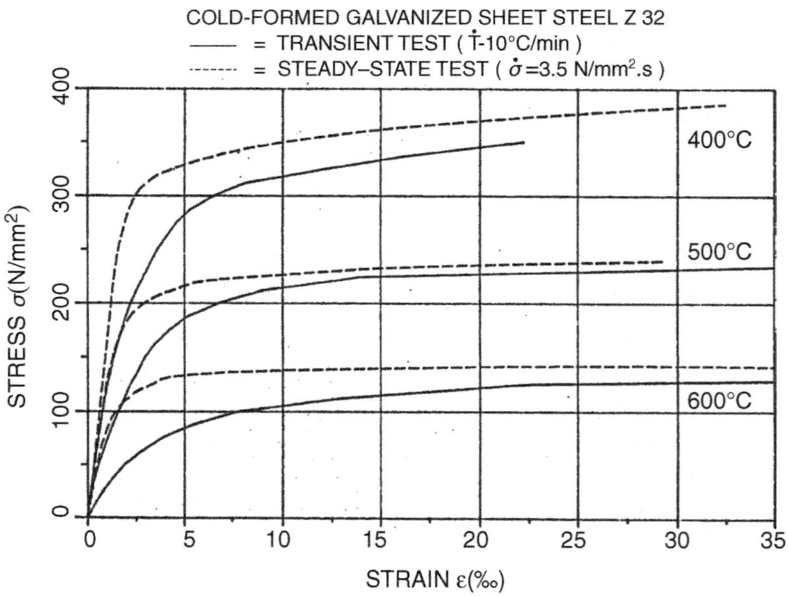

Figure 8-33. Stress–strain curves at 400°C, 500°C, and 600°C from steady-state and transient-state tests.
Source: Mäkeläinen and Miller (1983).

Table 8-6. Effective Yield Strength Retention Factors for the Cold-Formed Steel Stress–Strain–Temperature Model.

Steel Temperature [°C]	20	100	200	300	400	500	600	700	800	900	1000	1100	1200
Yield Strength Retention Factor	1.00	1.00	1.00	1.00	0.87	0.59	0.39	0.12	0.09	0.06	0.04	0.02	0.00

Source: ECCS (2001).

This model follows the popular Eurocode 3 (CEN 2006) stress–strain–temperature curves for structural steel in terms of the retention of modulus of elasticity and proportional limits at elevated temperatures. However, it specifies lower effective yield strength retention factors for CFS, as shown in Table 8-6. Currently, this remains the only stress–strain–temperature model for CFS that has been adopted by a code or standard.

8.2.4.1.1. Steady-State Response. The earliest CFS tensile coupon tests with application to SFED were undertaken by Klippstein (1980). Ten steady-state tests were carried out at a strain rate of 0.005/min on 1.2-mm-thick ASTM A446 Grade C (with a minimum specified yield strength of 275 MPa) specimens. The results were reported in terms of stress–strain curves, modulus of elasticity, and 0.2% offset yield strength for each temperature level.

The study by Mäkeläinen and Miller (1983) was more comprehensive, involving about 50 steady-state tests on 0.7-mm-thick grade Z32 (minimum specified yield strength of 320 MPa) galvanized sheet steel specimens at high (3.5 MPa/s) and low (0.35 MPa/s) loading rates. Stress–strain curves, modulus of elasticity, and 0.2% offset yield strength were reported for ten temperature levels up to 800°C. Figure 8-34 compares the stress–strain curves obtained at high and low loading rates.

More recently, steady-state experiments were reported for higher-strength steel grades G500 and G550 in China (Chen and Ye 2012) and for a broader range of CFS grades G250, G450, G500, and G550 in Australia (Ranawaka and Mahendran 2009, Kankanamge 2011).

Batista-Abreu (2015) described twenty-five elevated-temperature steady-state tests at temperature levels up 700°C. The experiments involved 18 ASTM A653 specimens tested at a strain rate of 0.003/min and seven ZAR345 (similar to ASTM A653) specimens tested at a strain rate of 0.005/min. All the specimens had specified minimum strengths of 230 MPa and 345 MPa and ranged in thickness from 1.15 mm to 2.58 mm. The results were presented in the form of measured stress–strain curves and retention

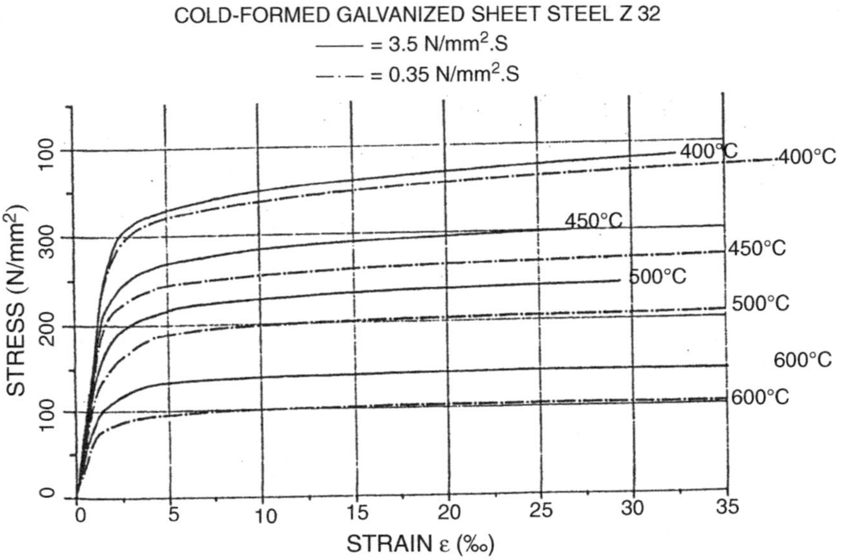

Figure 8-34. Stress–strain curves at 400°C, 450°C, 500°C, and 600°C from steady-state tests at high and low loading rates.
Source: Mäkeläinen and Miller (1983).

factors for modulus of elasticity, 0.2% offset yield strength, and ultimate strength.

8.2.4.1.2. Transient-State Response. In addition to steady-state tests described in the previous section, Mäkeläinen and Miller (1983) conducted transient-state tests with a heating rate of 10°C/min on 24 specimens (similar to those in the steady-state tests) at 16 stress levels ranging from 12.5 MPa to 350 MPa. The resulting temperature–strain curves were converted to stress–strain curves at nine temperature levels up to 600°C. Modulus of elasticity, 0.2% offset yield strength, and effective yield strength values derived from these transient test stress–strain curves were also reported for temperatures up 600°C and compared with corresponding values from steady-state tests.

Another comprehensive set of 88 transient-state tests at a 10°C/min heating rate was carried out by Sidey and Teague (1988). It involved four galvanized sheet steel grades with specified minimum yield strengths ranging from 220 MPa to 350 MPa and thicknesses ranging from 0.35 mm to 3.15 mm. The measured temperature–strain responses at stress levels ranging between 0.1 and 0.9 of the corresponding yield strength were reported and further converted to stress–strain curves for temperature levels up to 600°C.

The results indicated that sheet steel strength at elevated temperatures deteriorated to a greater degree than comparable steel grades for hot-rolled structural shapes. This trend was confirmed in later studies (Zhao et al. 2005, Outinen and Mäkeläinen 2007). Statistical treatment of the results also generated recommendations for CFS strength reduction factors at 0.5%, 1.5%, and 2% strains that were adopted by BS 5850 Part 8 (Lawson and Newman 1990, Lawson et al. 1993).

More recently, high-strength CFS grades G500 (with specified minimum strengths of 500 MPa) and G550 (with specified minimum strengths of 550 MPa) were investigated with transient tests in China (Chen and Ye 2012).

8.2.4.2 Residual Properties. Sidey and Teague (1988) carried out over 70 tensile coupon tests on specimens that were previously heated and held for 30 min at 600°C and then slowly cooled to ambient temperature. Based on these tests, the residual strength of CFS after fire exposure was permitted to be taken as 90% of the minimum specified strength in BS 5850 Part 8 (Lawson and Newman 1990)

Tests reported by Outinen and Mäkeläinen (2007) on specimens cut from CFS wall studs (with a specified minimum strength of 350 MPa) previously exposed to temperatures up to 950°C indicated that the BS 5850 recommendation was conservative (see Figure 8-35).

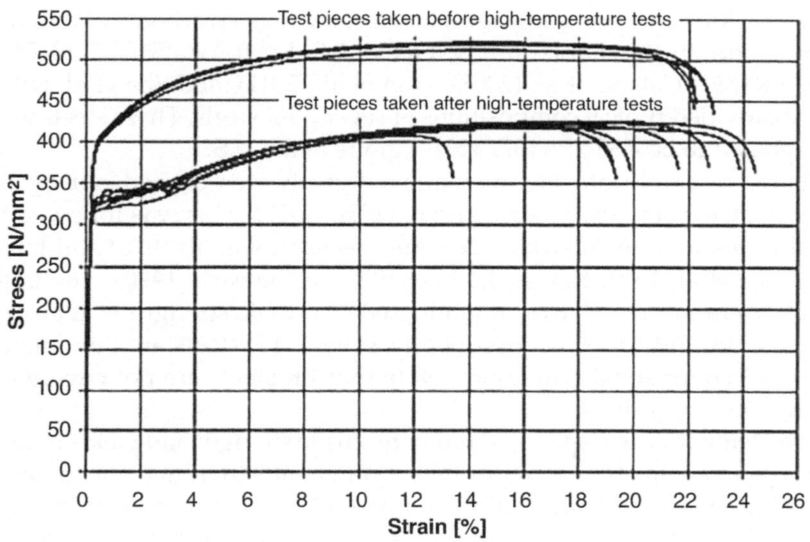

Figure 8-35. Tensile test results for cold-formed steel grade S350GD+Z before and after high-temperature exposure.
Source: Outinen and Mäkeläinen (2007).

8.2.5 Steel Grades with Improved Properties

8.2.5.1 Fire Resistant Steel Grades. Fire resistant (FR) steel grades have been in existence for some time and are used in both Europe and Japan. One of the most important characteristics of most FR steels is that their yield strength at 600°C equals or exceeds two-thirds of their ambient-temperature yield strength. In most cases, the elevated-temperature strength of these steels is achieved by adding proper amounts of alloying elements. However, this comes at an increased cost, which limits its wider adoption in building materials. ASTM Standard A1077/A1077M (2012) provides the specification for the use of structural steel with improved properties at high temperatures.

Among FR steel grades, the most common steels are those that are microalloyed with certain alloying elements. The increased strength at elevated temperatures is facilitated by the addition of proper amounts of alloying elements such as Mo, Cr, Nb, V, Ti, and so on. Among these elements, Mo has been found to be the most effective. The elevated-temperature strengths of these steels increase with an increase in Mo content. Most grades will typically have Mo in excess of 0.5 wt%.

The elevated-temperature strengthening in microalloyed FR steels is attributed to bainite strengthening and solid-solution strengthening. Solid-solution strengthening is the predominant elevated-temperature strengthening mechanism of Mo in FR steel. However, it is preferable to keep the Mo content at a low level to offset the high cost. Some of the common grades in the market are NSFR400, NSFR490, WGJ510C2, FRT800, and FRT850. Calado et al. (2008), Wan et al. (2011), and Sha et al. (1999) have provided typical compositions of several FR steels. These steels were designated to be ASTM A1077 (2012) grade in the U.S.

Early research into FR microalloyed steels was carried out by the French steel company Creusot-Loire (1976, 1977). This was followed by further research in Australia (Assefpour-Dezfuly et al. 1990) and Japan (Chijiwa et al. 1995, Kawasaki Steel 1993) in the early 1990s. The latter steels were more effective, retaining 60% of their strength at temperatures of around 650°C. Consequently, several FR steels are now manufactured commercially in Japan. Note that FR steels are not covered in AISC 360 Appendix 4.

The temperature-dependent ultimate tensile strength and yield strength for typical FR steels were reported by Wan et al. (2011) and Calado et al. (2008). In addition, Bailey (2016) compared the mechanical properties of several different FR steels whose properties were originally reported by Fujino et al. (1993), Sakumoto et al. (1992), Sakumoto and Saito (1995), and Kelly and Sha (1999). Figure 8-36 compares the reduction factors of yield strength (i.e., 0.2% strain offset from the temperature–stress–strain curves), and Figure 8-37 compares the reduction factors of tensile strength for vari-

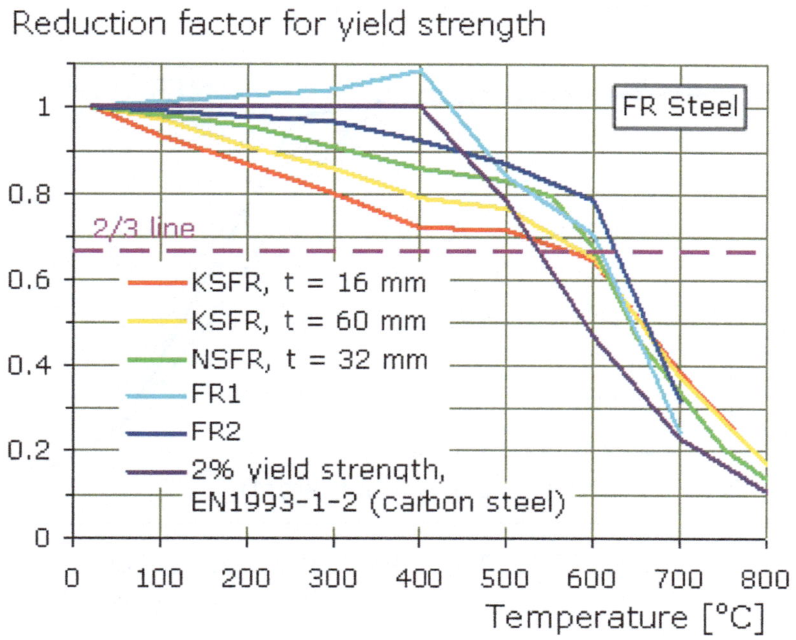

Figure 8-36. Strength reduction factors for yield strength of FR steel at elevated temperatures.
Source: Bailey (2016).

ous FR steels. These FR steels include Grade SM490B by Kawasaki Steel (denoted as KSFR) and Grade SM490A by Nippon Steel (denoted as NSFR), as well as two FR steel samples of Nippon Steel (denoted as FR1 and FR2, respectively). Although there are significant discrepancies between the test results, all FR steels appear to possess or exceed two-thirds of their ambient-temperature yield strength at 600°C.

Changes in elongation and tensile/yield strength as functions of temperature and strain rates for two FR steels were provided in Calado et al. (2008). Rothman (1988) provided detailed temperature-dependent data for Steel 17-22A (1Cr-1/Mo-V), which is an elevated-temperature service steel. Sakumoto et al. (1992) provided detailed temperature-dependent mechanical properties for a Japanese SM50A-NFR FR steel grade and compared them to a conventional structural steel grade. Seif et al. (2016a) [based on Luecke et al. (2011)] studied the elevated-temperature properties of two FR steels and compared their mechanical properties against ordinary structural steel. Figure 8-38 shows a comparison of normalized retained yield strength compared to the Eurocode model. Equation 2.4 in Figure 8-38 takes the following form (Seif et al. 2016a):

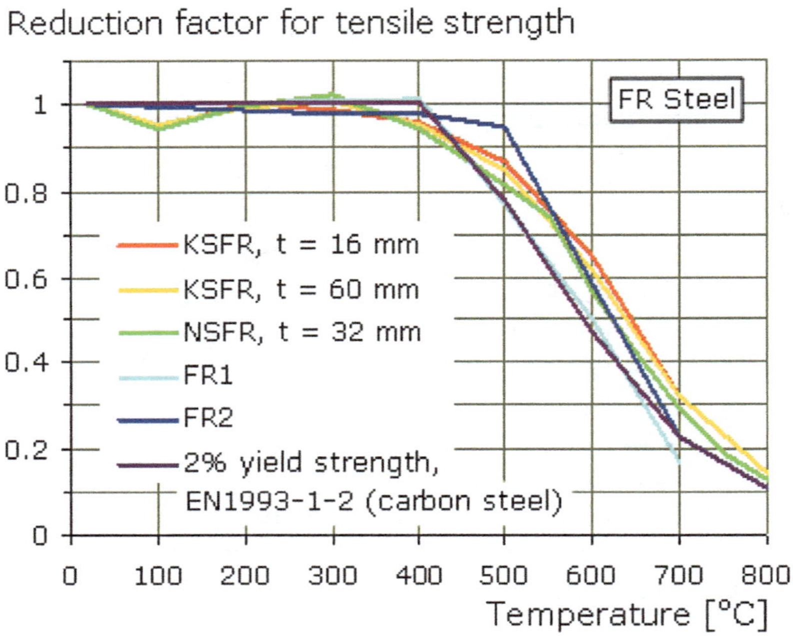

Figure 8-37. Strength reduction factors for tensile strength of FR steels at elevated temperatures.
Source: Bailey (2016).

$$R = \frac{F_{y0}}{S_y^0} = r_5 + (1 - r_5)\exp(-\frac{1}{2}\left(\frac{T^*}{r_3}\right)^{r_1} - \frac{1}{2}\left(\frac{T^*}{r_4}\right)^{r_2} \tag{8-1}$$

The dash/dot line for Equation 2.4 in Figure 8-38 is from nonlinear regression of the equation to the yield strength data for two FR steels. The regression set one parameter, r_2, at 1.0 (not fitted). Table 8-7 shows the values of the different parameters.

To obtain stress–strain data for FR steels as a function of temperature, the following equations are recommended (Seif et al. 2016a):

$$\text{for } \varepsilon < \frac{F_{y0}}{E}, \text{elastic behavior}, \sigma = E\varepsilon$$

$$\text{for } \varepsilon \geq \frac{F_{y0}}{E}, \text{plastic behavior},$$

$$\sigma = \left(RS_y^0 + \left(k_3 - k_4 S_y^0\right)\exp\left(-\left(\frac{T}{k_2}\right)^{k_1}\right)\left(\varepsilon - \frac{RS_y^0}{E}\right)^n\right)\left(\frac{\dot{\varepsilon}}{\dot{\varepsilon}_0}\right)^m \tag{8-2}$$

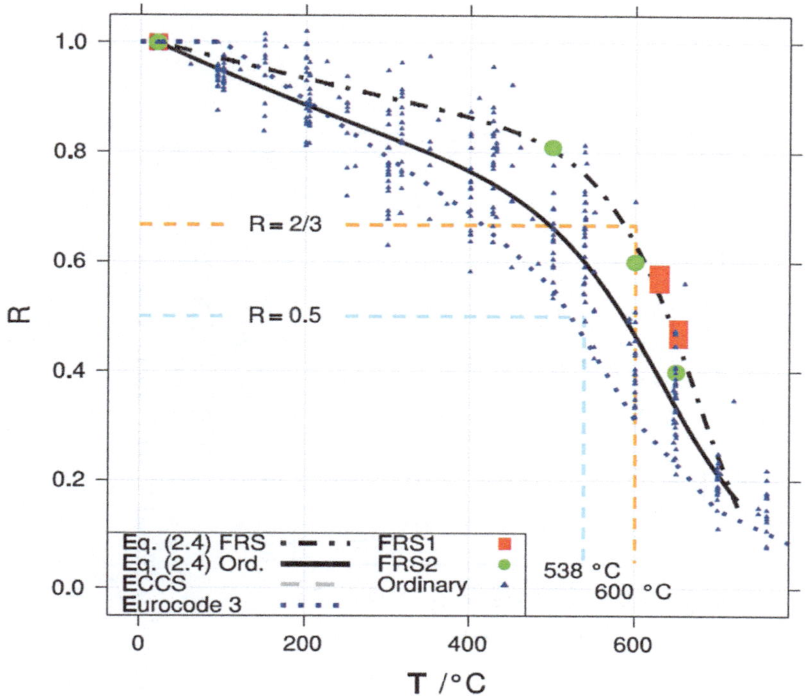

Figure 8-38. A comparison of normalized retained yield strength for ordinary structural steel and FR steels.
Source: Seif et al. (2016a).

Table 8-7. Parameter Values Used in the Stress–Strain Model and Yield Strength Equation.

Parameter	r_1	r_2	r_3	r_1	r_5	k_1	k_2	k_3	k_4	n
Value	9.782	1	625	1334	0	9.814	616	5835	15.846	0.456

Source: Seif et al. (2016a).

Using the values of the parameters in Table 8-7, the stress–strain data can be generated. Figure 8-39 shows temperature-dependent stress–strain data from the model by Seif et al. (2016a) in comparison to data for the two FR steels. Note that the values of k_i and n were fitted, in addition to the retained yield strength $R(T)$, to generate the model curves shown in Figure 8-39.

Additional data including creep behavior for a variety of FR steels were provided by Rothman (1988). It is recommended that the designer use

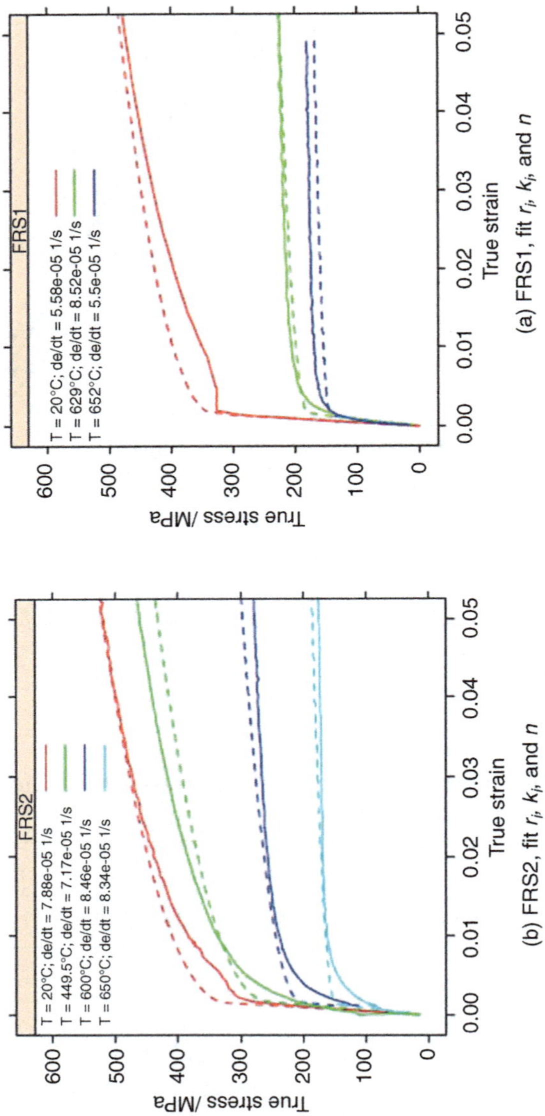

Figure 8-39. Stress–strain data (solid lines) and model plots (dotted lines) for two FR steels. Source: Seif et al. (2016a).

mechanical property data for an FR steel whose composition is close to one for which data are available in Rothman (1988). For the creep strain–time relationship of FR steel, the designer should refer to Fujimoto et al. (1980).

8.2.5.2 Stainless Steel Grades. Stainless steel is known for its corrosion resistance and aesthetic appearance, as well as ease of maintenance and construction. Structural application of stainless steel has been observed in buildings and bridges. Design specifications for stainless steel at elevated temperatures are available, such as the Australian Standard (AS) 4100 (1998), British Standard (BS) 5950 Part 8 (1998), and Eurocode 3 (CEN 2006). The reduction factors for the mechanical properties of stainless steel at elevated temperatures are provided in the following section, granting that they are based on data obtained for hot-rolled carbon steel in most cases.

The material properties of stainless steel are significantly different from those of carbon steel. Hence, property data for carbon steel may not be used for stainless steel without appropriate changes. For carbon and low-alloy steels, the proportional limit is assumed to be at least 70% of the yield point, but for stainless steel the proportional limit ranges from approximately 36% to 60% of the yield strength. These properties can be determined by conducting controlled tests or using data provided in the following sections if the steel grades are the same.

Many stainless steels, particularly certain austenitic and precipitation-hardened steels, are used extensively for elevated-temperature applications. Stainless steel grades that are increasingly being used are AISI 304, AISI 316, and EN1.4462. The following section discusses mechanical property data for AISI 304 and EN1.4462 grades. Data for AISI 316 steel can be found in Rothman (1988). Also, Zeng et al. (2013) provide details on the usage of AISI 316 steel as concrete reinforcing bars.

8.2.5.2.1. EN1.4462 (Duplex) Steel and EN 1.4301 (AISI 304) Steel. Chen and Young (2006) conducted experiments to study the behavior of EN 1.4462 and EN 1.4301 at both ambient and elevated temperatures. Table 8-8 shows

Table 8-8. Mechanical Properties of Stainless Steel Types EN 1.4462 and EN 1.4301 at Ambient Temperature.

Stainless steel	$f_{0.2.\text{normal}}$ (MPa)	$f_{0.5.\text{normal}}$ (MPa)	$f_{1.5.\text{normal}}$ (MPa)	$f_{2.0.\text{normal}}$ (MPa)	$f_{u.\text{normal}}$ (MPa)	$\varepsilon_{u.\text{normal}}$ (%)	E_{normal} (GPa)
EN 1.4462 (Duplex)	731	729	817	824	870	15.8	227
EN 1.4301 (AISI 304)	398	409	443	452	709	60.6	187

Source: Chen and Young (2006).

the mechanical properties for the EN1.4462 and EN1.4301 stainless steels measured at room temperature. Because the stress–strain behavior of stainless steel does not show a sharp yield plateau, results for the yield stress of these stainless steels are reported based on different definitions of the yield stress. Specifically, $f_{0.2}$ represents a yield stress corresponding to the 0.2% offset strain, whereas $f_{0.5}$, $f_{1.5}$, and $f_{2.0}$ represent yield stresses corresponding to 0.5%, 1.5%, and 2.0% total strains, respectively. Results from elevated-temperature tests on EN1.4462 and EN1.4301 stainless steels in terms of retention factors for the yield strength and elastic modulus are summarized in Table 8-9.

Figure 8-40 shows a comparison between computed, measured, and literature data on yield strength reduction factors.

The proposed equation for ultimate strength is given as follows (Chen and Young 2006):

$$\frac{f_{u,T}}{f_{u,\text{normal}}} = a - \frac{(T-b)^n}{c} \tag{8-3}$$

The coefficients in Equation (8-3) are obtained from Table 8-10.

Based on the elevated-temperature tests, Chen and Young (2006) proposed a stress–strain model for stainless steels at elevated temperatures. That model comprised the following equations:

$$\varepsilon_T = \begin{cases} \dfrac{f_T}{E_T} + 0.002\left(\dfrac{f_T}{f_{y,T}}\right)^{n_T} & \text{for } f_T \le f_{y,T} \\[2ex] \dfrac{f_T - f_{y,T}}{E_{y,T}} + \varepsilon_{u,T}\left(\dfrac{f_T - f_{y,T}}{f_{u,T} - f_{y,T}}\right)^{m_T} + \varepsilon_{y,T} & \text{for } f_T > f_{y,T} \end{cases}$$

and

$$E_{y,T} = \frac{E_T}{1 + 0.002 n_T E_T / f_{y,T}}$$

$$n_T = 6 + 0.2\sqrt{T}$$

$$m_T = \begin{cases} 5.6 - \dfrac{T}{200} & \text{for stainless steel type EN 1.4462} \\[2ex] 2.3 - \dfrac{T}{1000} & \text{for stainless steel type EN 1.4301} \end{cases} \tag{8-4}$$

The temperature-dependent creep behavior of AISI 304 steels is given in Table 8-11.

Table 8-9. Reduction Factors for Yield Strength and Elastic Modulus at Elevated Temperature of Stainless Steel Types EN 1.4462 and EN 1.4301.

	Temperature (°C)	22	80	180	320	450	550	660	760	870	960
EN 1.4462 (Duplex)	$f_{0.2.T}/f_{0.2.normal}$	1.000	0.943,0.923[a]	0.834	0.758	0.728	0.683	0.528	0.354	0.170	0.031, 0.038[a]
	$f_{0.5.T}/f_{0.5.normal}$	1.000	0.936, 0.920[a]	0.834	0.742	0.717	0.678	0.539	0.351	0.177	0.033, 0.041[a]
	$f_{1.5.T}/f_{1.5.normal}$	1.000	0.949, 0.919[a]	0.834	0.802	0.752	0.718	0.542	0.324	0.163	0.031, 0.037[a]
	$f_{2.0.T}/f_{2.0.normal}$	1.000	0.947, 0.922[a]	0.836	0.814	0.765	0.729	0.541	0.325	0.161	0.032, 0.036[a]
	E_T/E_{normal}	1.000	0.947, 0.949[a]	0.863	0.714	0.696	0.690	0.643	0.409	0.300	0.060, 0.063[a]
	Temperature (°C)	22	80	180	320	450	550	660	760	870	960
EN 1.4301 (AISI 304)	$f_{0.2.T}/f_{0.2.normal}$	1.000	0.839	0.764	0.698	0.678	0.595	0.523	0.357	0.181	0.116, 0.101[a]
	$f_{0.5.T}/f_{0.5.normal}$	1.000	0.848	0.773	0.704	0.687	0.599	0.543	0.359	0.176	0.115, 0.098[a]
	$f_{1.5.T}/f_{1.5.normal}$	1.000	0.860	0.786	0.720	0.693	0.619	0.562	0.357	0.163	0.106, 0.088[a]
	$f_{2.0.T}/f_{2.0.normal}$	1.000	0.863	0.788	0.730	0.704	0.635	0.577	0.347	0.159	0.104, 0.089[a]
	E_T/E_{normal}	1.000	1.045	0.939	1.036	0.951	0.900	0.862	0.574	0.452	0.340, 0.351[a]

[a]Second test.

Source: Chen and Young (2006).

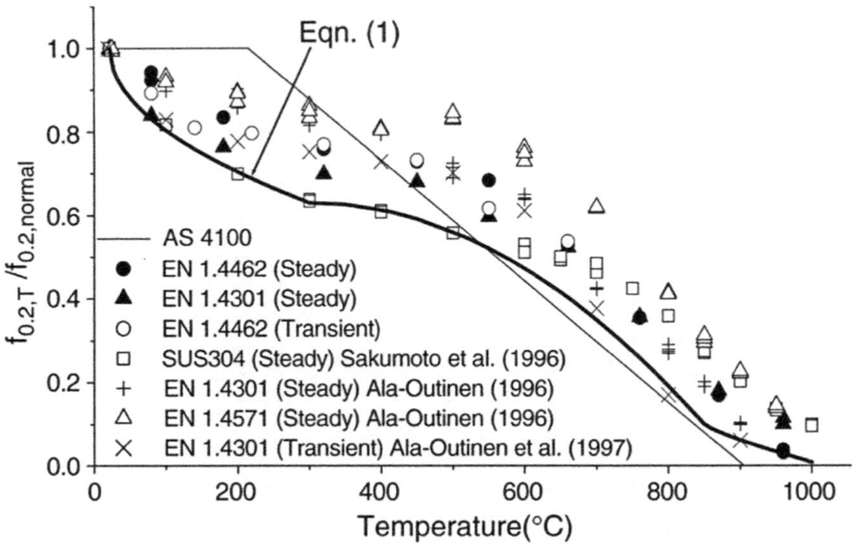

Figure 8-40. Comparison of reduction factor of 0.2% yield strength predicted by AS 4100 and proposed equation with test results.
Source: Chen and Young (2006).

8.2.5.2.2. Stress–Strain–Temperature Relationships in Design Codes. The stress–strain relationship provided by Eurocode 3 (CEN 2006) for stainless steel at elevated temperatures is applicable for heating rates between 2 K/min and 50 K/min. The relevant equations are shown in Figure 8-41.

where
$f_{u,\theta}$ = tensile strength
$f_{0.2p,\theta}$ = proof strength at 0.2% plastic strain;
$E_{a,\theta}$ = slope of the linear elastic range;
$E_{ct,\theta}$ = slope at proof strength;
$\varepsilon_{c,\theta}$ = total strain at proof strength; and
$\varepsilon_{u,\theta}$ = ultimate strain.

Table 8-12 provides reduction factors (defined in the following list) relative to the appropriate value at 20°C for the stress–strain relationship of EN1.4301 stainless steel at elevated temperatures as follows:

- Slope of linear elastic range, relative to slope at 20°C: $k_{E,\theta}=E_{a,\theta}\ /\ E_a$,
- Proof strength, relative to yield strength at 20°C: $k_{0.2p,\theta}=f_{0.2p,\theta}\ /\ f_y$,
- Tensile strength, relative to tensile strength at 20°C: $k_{u,\theta}=f_{u,\theta}/f_u$, and
- Correction factor for yield strength: $k_{2\%,\theta}$.

Table 8-10. Coefficients of the Proposed Equation for Ultimate Strength.

Ultimate strength Eq. (3)	EN 1.4462 (Duplex)	Temperature (°C)	$22 \leq T < 450$	$450 \leq T < 660$	$660 \leq T \leq 960$
		a	0.85	0.85	0.51
		b	450	450	660
		c	9.6×10^{13}	1.3×10^5	200
		n	5	2	0.8
	EN 1.4301 (AISI 304)	Temperature (°C)	$22 \leq T < 450$	$450 \leq T < 660$	$660 \leq T \leq 960$
		a	0.7	0.7	0.06
		b	450	450	960
		c	4.8×10^{13}	1.92×10^5	-2.2×10^5
		n	5	2	2

Source: Chen and Young (2006).

The correction factor for the yield strength, $k_{2\%,\theta}$, is used for simple calculation methods. It is assumed that the *effective* yield strength to be used in simple calculation methods should be between the values of the proof strength $f_{0.2p,\theta}$ and the tensile strength f_u as given by

$$f_{y,\theta} = f_{0.2p,\theta} + k_{2\%,\theta}(f_{u,\theta} - f_{0.2p,\theta}) \tag{8-5}$$

Values for other grades with chemical compositions shown in Table 8-13 (Bailey 2017) are given in Eurocode 3 (CEN 2006).

Eurocode 3 (CEN 2006) assumes the same reduction rate for the slope of the linear elastic range $k_{E,\theta}$ for all steel grades.

8.3 MECHANICAL PROPERTIES OF CONCRETE AND REINFORCING STEEL

8.3.1 Concrete

Concrete is available in various forms. Accordingly, its mechanical properties, as well as its tendency for spalling, depend on the composition of the constituent material and its compressive strength. The strength of concrete influences its mechanical properties, as well as the tendency for spalling. A concrete with compressive strength in the range of 20 MPa to 50 MPa is classified as normal-strength concrete (NSC), and those falling in the range of 50 MPa to 120 MPa are classified as high-strength concrete (HSC).

Table 8-11. Stress Rupture and Creep Data for AISI 304 Steel.

TEMP C	RPSTR 1000 h MPa	RPSTR 1000 h ksi	RPSTR(a) 10000 h MPa	RPSTR(a) 10000 h ksi	RPSTR(a) 100000 h MPa	RPSTR(a) 100000 h ksi	Stress(b) 0.0001%/h MPa	Stress(b) 0.0001%/h ksi	Stress(b) 0.00001%/h MPa	Stress(b) 0.00001%/h ksi
480	338	49.0	262	38.0	...	...	...	...	...	...
540	241	35.0	183	26.5	145	21.0	138	20.0	83	12.0
595	152	22.0	117	17.0	90	13.0	90	13.0	52	7.5
650	97	14.0	69	10.0	48	7.0	57	8.3	30	4.3
705	59	8.5	41	6.0	28	4.0	34	5.0	17	2.5
760	41	6.0	26	3.7	17	2.5	21	3.0	10	1.5
815	26	3.7	19	2.7	10	1.5	17	2.4	8	1.2
870	19	2.7	...	...	...	...	...	...	...	...

(a) Extrapolated (b) Stress for secondary creep rate of extrapolated data
Source: Ref 15, p 275

Source: Harvey (1982).

Strain range	Stress σ	Tangent modulus E_t
$\varepsilon \leq \varepsilon_{c,\theta}$	$\dfrac{E \cdot \varepsilon}{1 + a \cdot \varepsilon^b}$	$\dfrac{E(1 + a \cdot \varepsilon^b - a \cdot b \cdot \varepsilon^b)}{(1 + a \cdot \varepsilon^b)^2}$
$\varepsilon_{c,\theta} < \varepsilon \leq \varepsilon_{u,\theta}$	$f_{0.2p,\theta} - e + (d/c)\sqrt{c^2 - (\varepsilon_{u,\theta} - \varepsilon)^2}$	AC2 $\dfrac{d \cdot (\varepsilon_{u,\theta} - \varepsilon)}{c\sqrt{c^2 - (\varepsilon_{u,\theta} - \varepsilon)^2}}$ AC2
Parameters	$\varepsilon_{c,\theta} = f_{0.2p,\theta}/E_{a,\theta} + 0.002$	
Functions	$a = \dfrac{E_{a,\theta}\,\varepsilon_{c,\theta} - f_{0.2p,\theta}}{f_{0.2p,\theta}\,\varepsilon_{c,\theta}^{\ b}}$ $c^2 = (\varepsilon_{u,\theta} - \varepsilon_{c,\theta})\left(\varepsilon_{u,\theta} - \varepsilon_{c,\theta} + \dfrac{e}{E_{ct,\theta}}\right)$ $e = \dfrac{(f_{u,\theta} - f_{0.2p,\theta})^2}{(\varepsilon_{u,\theta} - \varepsilon_{c,\theta})E_{ct,\theta} - 2(f_{u,\theta} - f_{0.2p,\theta})}$	$b = \dfrac{(1 - \varepsilon_{c,\theta}\,E_{ct,\theta} / f_{0.2p,\theta})\,E_{a,\theta}\,\varepsilon_{c,\theta}}{(E_{a,\theta}\,\varepsilon_{c,\theta} / f_{0.2p,\theta} - 1)\,f_{0.2p,\theta}}$ $d^2 = e(\varepsilon_{u,\theta} - \varepsilon_{c,\theta})E_{ct,\theta} + e^2$

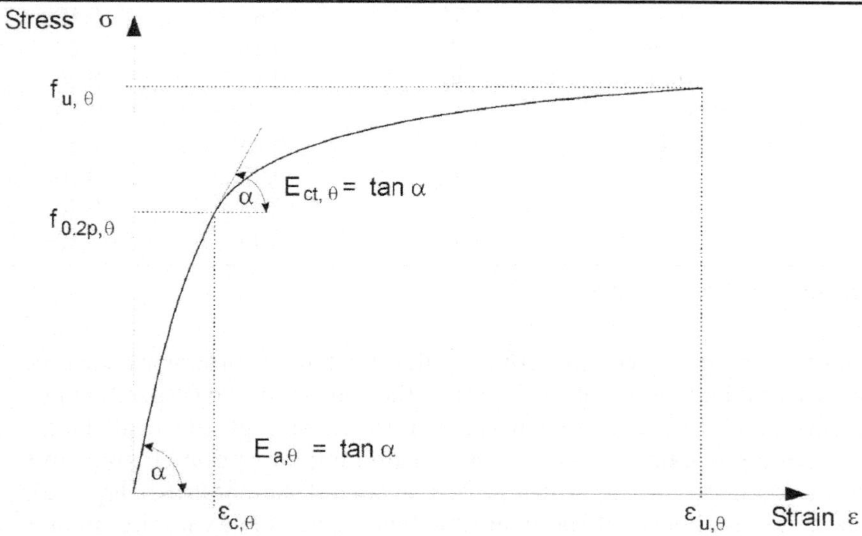

Key:

$f_{u,\theta}$	is	tensile strength;
$f_{0.2p,\theta}$	is	the proof strength at 0.2% plastic strain;
$E_{a,\theta}$	is	the slope of the linear elastic range;
$E_{ct,\theta}$	is	the slope at proof strength;
$\varepsilon_{c,\theta}$	is	the total strain at proof strength;
$\varepsilon_{u,\theta}$	is	the ultimate strain.

Figure 8-41. Stress–strain relationships and relevant equations. Source: Eurocode 3 (CEN 2006).

Table 8-12. Reduction Factors for EN1.4301 Stainless Steel.

Steel Temperature θ_a	Reduction factor (relative to E_a) for the slope of the linear elastic range $k_{E.\theta} = E_{a.\theta}/E_a$	Reduction factor (relative to f_y) for proof strength $k_{0\cdot 2p.\theta} = f_{0.2p.\theta}/f_y$	Reduction factor (relative to f_u) for tensile strength $k_{u.\theta} = f_{u.\theta}/f_u$	Factor for determination of the yield strength $f_{y.\theta}$ $k_{2\%.\theta}$
Grade 1.4301				
20	1,00	1,00	1,00	0,26
100	0,96	0,82	0,87	0,24
200	0,92	0,68	0,77	0,19
300	0,88	0,64	0,73	0,19
400	0,84	0,60	0,72	0,19
500	0,80	0,54	0,67	0,19
600	0,76	0,49	0,58	0,22
700	0,71	0,40	0,43	0,26
800	0,63	0,27	0,27	0,35
900	0,45	0,14	0,15	0,38
1000	0,20	0,06	0,07	0,40
1100	0,10	0,03	0,03	0,40
1200	0,00	0,00	0,00	0,40

Source: Eurocode 3 (CEN 2006).

8.3.1.1 Compressive Strength. Unlike mechanical property measurements at ambient temperature, in which the specimen size is specified per industry standards, measurements of compressive strength at elevated temperatures are usually carried out on a wide range of specimen sizes and loading conditions owing to the lack of relevant test standards (Khaliq 2012). Hence, the designer should be aware that there is inevitable variation among the available test data at elevated temperatures.

Figures 8-42 and 8-43 show the variation of the normalized compressive strength of NSC and HSC, respectively, as a function of temperature. It can be observed from Figure 8-42 that elevated temperatures have a minimal effect on the compressive strength of NSC up to 200°C. This is because of the highly permeable nature of NSC, which allows easy diffusion of pore pressure. Conversely, there is a more rapid rate of degradation in the compressive strength of HSC at elevated temperatures, as shown in Figure 8-43. This is because of the compact microstructure of HSC that arises owing to the inclusion of a binder material such as slag or silica fume. The resulting impermeability of HSC makes it susceptible to strength deterioration and

Table 8-13. Comparison of Stainless Steel Grades.

Steel designation		Chemical composition (% by mass)									
No.	name	C max.	Si max.	Mn max.	P max.	S max.	N	Cr	Mo	Ni	Others
1.4301	X5 CrNi 18-10	0.07	1.0	2.0	0.045	0.016	0.11	17.5 to 19.5		8.0 to 10.5	
1.4401	X5 CrNiMo 17-12-2	0.07	1.0	2.0	0.045	0.015	0.11	16.5 to 18.5	2.0 to 2.5	10.0 to 13.0	
1.4404	X2 CrNiMo 17-12-2	0.03	1.0	2.0	0.045	0.015	0.11	16.5 to 18.5	2.0 to 2.5	10.0 to 13.0	
1.4571	X6 CrNiMoTi 17-12-2	0.08	1.0	2.0	0.045	0.015		16.5 to 18.5	2.0 to 2.5	10.0 to 13.5	Ti:5xC to 0.70
1.4003	X2 CrNi 12	0.03	1.0	1.5	0.040	0.015	0.03	10.5 to 12.5		0.3 to 1.0	
1.4462	X2 CrNiMoN 22-5-3	0.03	1.0	2.0	0.035	0.015	0.10 to 0.22	21.0 to 23.0	2.5 to 3.5	4.5 to 6.5	

Source: Bailey (2007) as adapted from Eurocode 3 (CEN 2006).

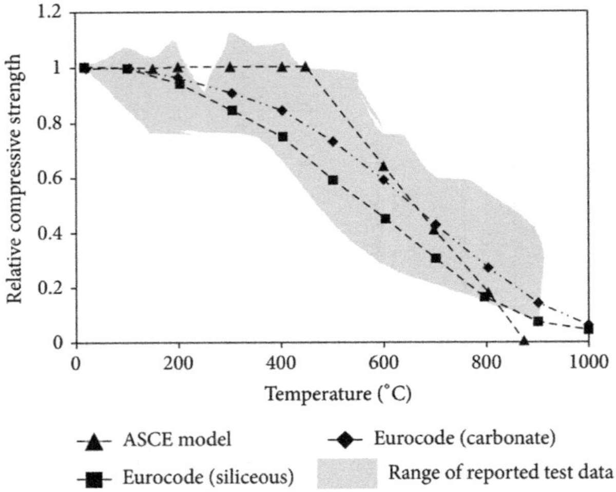

Figure 8-42. Relative variation of compressive strength of NSC as a function of temperature.
Source: Kodur (2014).

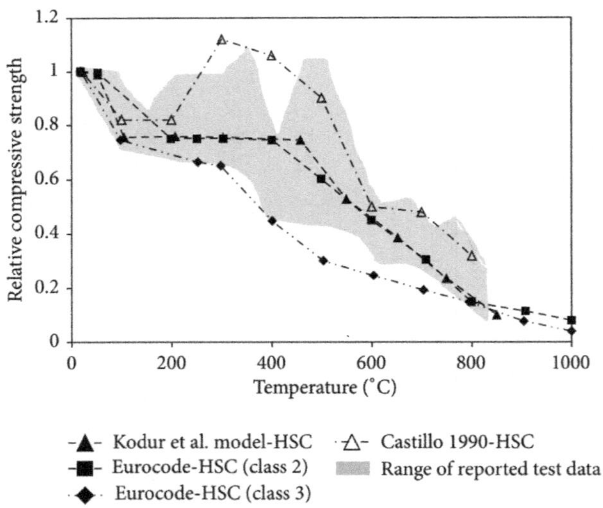

Figure 8-43. Relative variation of compressive strength of HSC as a function of temperature.
Source: Kodur (2014).

spalling. The variation among the data observed in both figures can be attributed to the difference in heating or loading rates, specimen size and curing, moisture content, age of the specimen, and presence of admixture (Kodur 2014).

8.3.1.2 Stress–Strain Response. Figures 8-44 and 8-45 plot the measured stress–strain responses of NSC and HSC, respectively. With increasing temperature, the peak compressive stress decreases as the corresponding strain increases for both cases. Also, the ductility and ultimate strain also increase with increasing temperature. Compared to NSC, HSC exhibits a more brittle response.

Design curves for the stress–strain–temperature response of concrete can be found in ASCE (1992) and in Eurocode 2 (CEN 2004). The stress–strain relationship for all types of aggregates according to ASCE (1992) is given by the following equations:

$$\text{For ascending branch}: \varepsilon_c \le \varepsilon_{max}, f_c = f_c'\left[1-\left(\frac{\varepsilon_{max}-\varepsilon_c}{\varepsilon_{max}}\right)^2\right] \tag{8-6}$$

$$\text{For descending branch}: \varepsilon_c \ge \varepsilon_{max}, f_c = f_c'\left[1-\left(\frac{\varepsilon_c-\varepsilon_{max}}{3\varepsilon_{max}}\right)^2\right] \tag{8-7}$$

where

$$f_c' = f_{co}', \ if\ T \le 450°C \tag{8-8}$$

$$f_c' = f_{co}'\left[2.011-2.353.\frac{T-20}{1000}\right], \ if\ T \ge 450°C \tag{8-9}$$

$$\varepsilon_{max} = 0.0025 + (6.0T + 0.04T^2) \times 10^{-6} \tag{8-10}$$

Per the preceding equations,
ε_c = strain in concrete,
ε_{max} = concrete strain at maximum stress,
f_c = stress in concrete (MPa),

f_c' = temperature-dependent concrete strength (MPa), and

f_c' = concrete strength at ambient (MPa).

Eurocode 2 (CEN 2004) provides a stress–strain curve for uniaxially stressed concrete at elevated temperatures, which is shown in Figure 8-46.

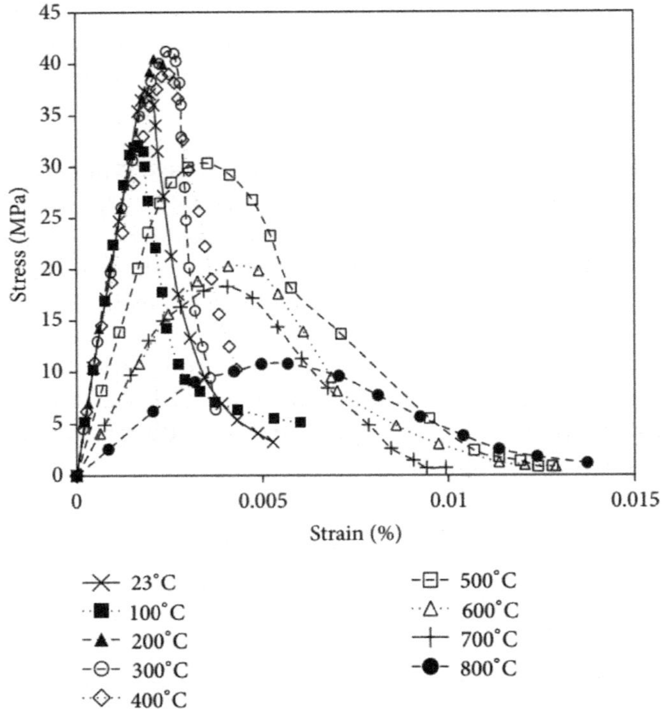

Figure 8-44. Stress–strain diagram of NSC at elevated temperatures.
Source: Kodur (2014).

The stress is related to strain using the following equation for the ascending and descending branches of the curve:

$$\sigma(\theta) = \frac{3\varepsilon f_{c,\theta}}{\varepsilon_{c1,\theta}\left(2+\left(\frac{\varepsilon}{\varepsilon_{c1,\theta}}\right)^3\right)}, \quad for\ \varepsilon \le \varepsilon_{c1,\theta} \tag{8-11}$$

Linear or nonlinear models are permitted:

$$for\ \varepsilon_{c1,\theta} \le \varepsilon \le \varepsilon_{cu1,\theta} \tag{8-12}$$

The ascending branch is defined by two parameters, the compressive strength $f_{c,\theta}$ and the strain $\varepsilon_{c1,\theta}$ corresponding to $f_{c,\theta}$. These parameters are temperature-dependent and can be found using Table 8-14 per Eurocode 2 (CEN 2004).

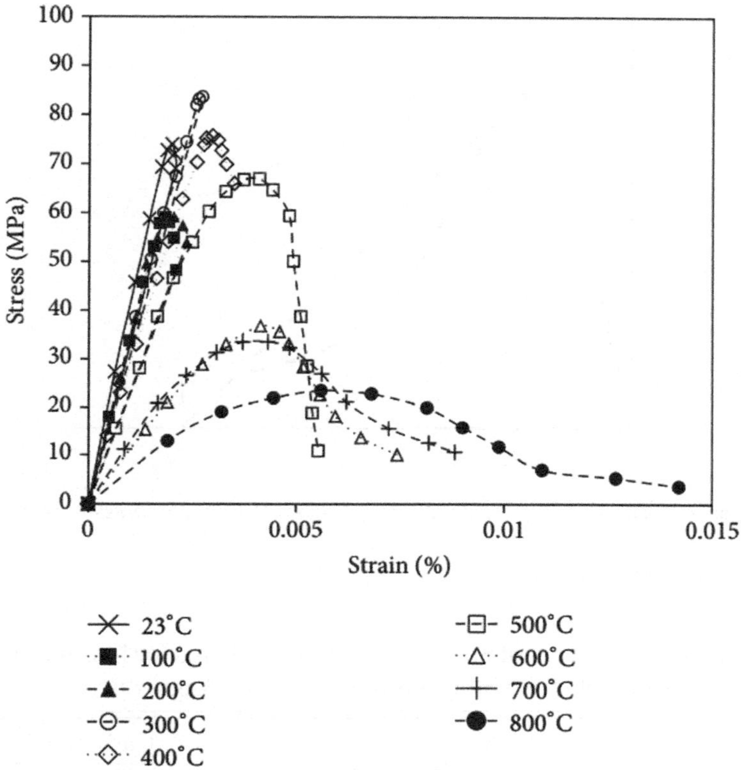

Figure 8-45. Stress–strain diagram of HSC at elevated temperatures. Source: Kodur (2014).

8.3.1.3 Tensile Strength. Figure 8-47 plots the tensile strengths of HSC and NSC as functions of temperature based on various test data (Behnood and Ghandehari 2009; Carette et al. 1982; Felicetti et al. 1996; CEN 2004). The tensile strength of NSC and HSC decreases at elevated temperatures. At 300°C, NSC loses approximately 20% of its tensile strength. Above 300°C, the tensile strength of NSC decreases at a rapid rate. This is due to more pronounced thermally induced damage in the form of microcracks (Kodur 2014). The designer should be aware that there is inevitable variation among the available test data.

8.3.1.4 Elastic Modulus. Figure 8-48 plots the elastic moduli of HSC and NSC as functions of temperature based on test data (Castillo and Durrani 1990, Phan 1996, CEN 2004). The elastic moduli of HSC and NSC decrease with increasing temperature. The designer should be aware that there is inevitable variation among the available test data. Hence, the designer

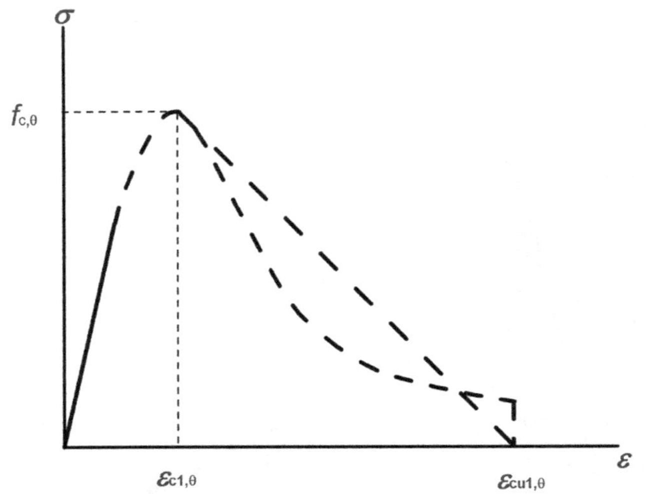

Figure 8-46. Stress–strain relationships of concrete under compression at elevated temperatures.
Source: Eurocode 2 (CEN 2004).

Table 8-14. Parameters of Stress–Strain Relationships for Normal-weight Concrete with Siliceous or Calcareous Aggregates.

	Siliceous			Carbonate		
Temperature, θ (°C)	$\frac{f_{c,\theta}}{f_{ck}}$	$\varepsilon_{c1,\theta}$	$\varepsilon_{cu1,\theta}$	$\frac{f_{c,\theta}}{f_{ck}}$	$\varepsilon_{c1,\theta}$	$\varepsilon_{cu1,\theta}$
20	1.00	0.0025	0.0200	1.00	0.0025	0.0200
100	1.00	0.004	0.0225	1.00	0.0040	0.0225
200	0.95	0.0055	0.0250	0.97	0.0055	0.0250
300	0.85	0.0070	0.0275	0.91	0.0070	0.0275
400	0.75	0.0100	0.0300	0.85	0.0100	0.0300
500	0.60	0.0150	0.0325	0.74	0.0150	0.0325
600	0.45	0.0250	0.0350	0.60	0.0250	0.0350
700	0.30	0.0250	0.0375	0.43	0.0250	0.0375
800	0.15	0.0250	0.0400	0.27	0.0250	0.0400
900	0.08	0.0250	0.0425	0.15	0.0250	0.0425
1000	0.04	0.0250	0.0450	0.06	0.0250	0.0450
1100	0.01	0.0250	0.0475	0.02	0.0250	0.0475
1200	0.00	–	–	0.00	–	–

$f_{c,\theta}$ is the characteristic cylinder strength (MPa)
$\varepsilon_{cu1,\theta}$ is the ultimate strain
Source: Eurocode 2 (CEN 2004).

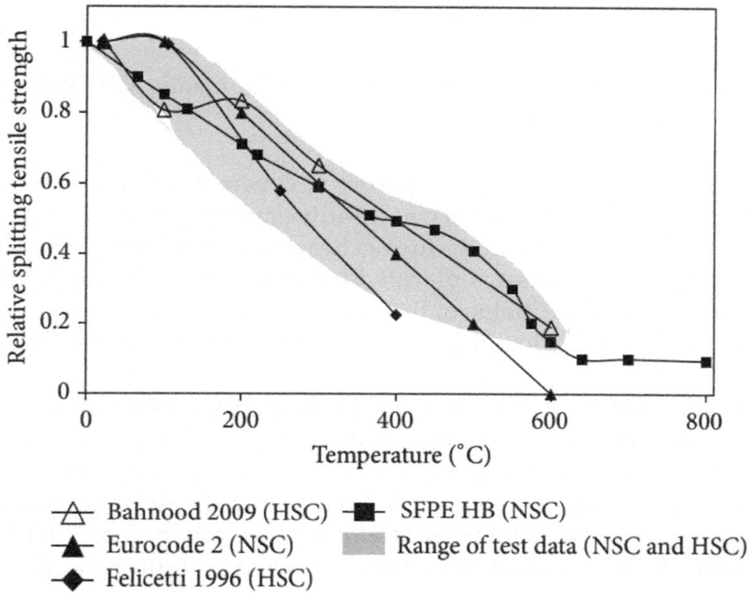

Figure 8-47. Relative variations of tensile strength of HSC as a function of temperature.
Source: Kodur (2014).

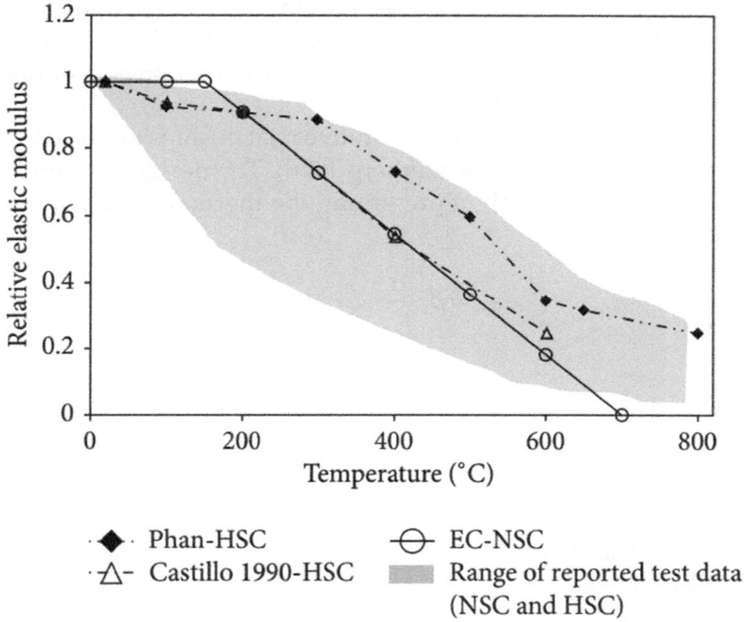

Figure 8-48. Relative variations of modulus of elasticity of concrete as a function of temperature.
Source: Kodur (2014).

should use the lower-bound values unless specific justification is provided for adopting other curves.

8.3.1.5 Creep. The creep properties of concrete depend on the type of aggregate used, the chemical composition, and the chemical and physical reactions in the concrete during heating (Schneider 1988). Creep strain becomes significant at elevated temperatures owing to the rapid movement of moisture. Some of the factors that affect creep are temperature, stress level, time, loading, and mix design of the concrete (Dwaikat 2009).

When concrete is exposed to elevated temperatures, its moisture content and cement paste chemical composition undergo complex changes. Also, there is differential thermal expansion between the cement paste and the aggregate. These changes collectively lead to internal stresses and microcracking, which result in transient strains in the concrete (Schneider 1988). Transient strain in concrete develops during its first exposure to heating only (Khoury et al. 1985).

There is no direct test method to measure transient strain. Transient and creep strain are quantified together by measuring the total strain and deducting both the mechanical and the free thermal strain. This lumped quantity is referred to as transient creep strain. Common references for concrete properties at elevated temperatures do not provide a direct relationship either for transient strain or transient creep strain (ASCE 1992, CEN 2004). Rather, these effects are included implicitly in their stress–strain relationship by providing an allowance (Khaliq 2012).

The equations that follow are constitutive relationships for the creep and transient strain of concrete (Anderberg 1976). The designer should exercise caution when explicitly representing the thermal creep behavior of concrete.

$$\varepsilon_{cr} = \beta_1 \frac{\sigma}{f_{c,T}} \sqrt{t} e^{d(T-293)} \tag{8-13}$$

$$\varepsilon_{tr} = k_2 \frac{\sigma}{f_{c,20}} \varepsilon_{th} \tag{8-14}$$

where

ε_{cr} = creep strain

ε_{tr} = transient strain

$\beta_1 = 6.28 \times 10^{-6}\ s^{-0.5}$

$d = 2.658 \times 10^{-3}\ K^{-1}$

T = concrete temperature (K) at time t (s)

$f_{c,}$ = concrete strength at temperature T

σ = stress in the concrete at the current temperature

k_2 = a constant between 1.8 and 2.35
ε_{th} = thermal strain
$f_{c,20}$ = concrete strength at room temperature

8.3.1.6 Spalling. Fire-induced spalling in concrete is characterized by two broad theories (Kodur 2000). The first theory is based on the pore-pressure buildup that occurs during heating of concrete. The pressure increase is higher in HSC as compared to NSC owing to its comparatively low permeability and high density. Consequently, HSC is more susceptible to spalling as compared to NSC because the water vapor pressure generated during heating cannot escape. Figure 8-49 illustrates the spalling of concrete when the effective pore pressure, which is a product of porosity and pore pressure, exceeds the tensile strength of the concrete.

The second theory for characterizing concrete spalling considers restrained thermal dilation close to the heated surface, which leads to the development of compressive stress parallel to the heated surface (as illustrated in Figure 8-50). These thermal stresses may lead to subsurface cracks and may be released by spalling or brittle fracture of the concrete. Overall, there is insufficient research for a designer to quantitatively predict the occurrence of spalling. Hence, designers should rely on practical mitigation of spalling if deemed pertinent (e.g., use of HSC). Practical mitigation would include the following: use of polypropylene microfibers, additional air entrainment, thicker sections, and use of low-expansion aggregates. Also, the designer may follow mitigation recommendations from Eurocode 2 (CEN 2004).

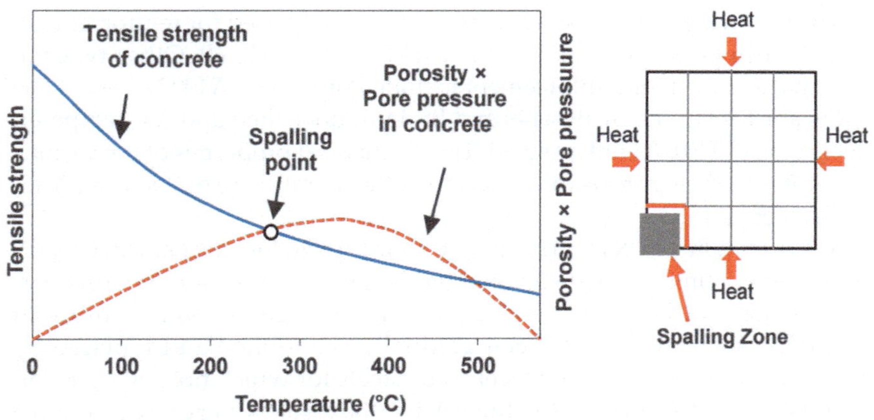

Figure 8-49. Illustration of spalling of concrete due to pressure buildup. Source: Dwaikat (2009).

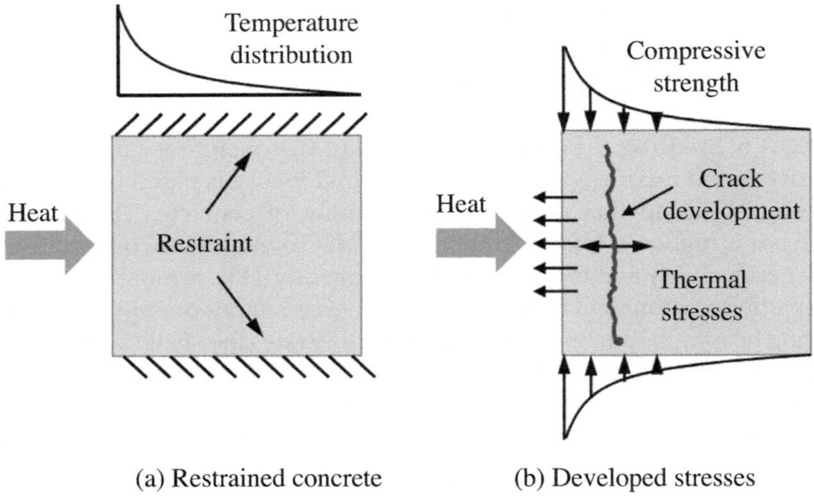

Figure 8-50. Illustration of spalling of concrete due to thermal dilatation. Source: Khaliq (2012).

8.3.2 Reinforcing Steel

Steel reinforcement includes the following: carbon steel bars (plain and deformed), cold reduced plain and deformed steel wire, stainless steel bars and fabric, and galvanized carbon steel bars and fabrics. Carbon steel bars include plain round mild steel bars and deformed high-yield steel bars. Both hot rolling and cold working are used to produce high-yield bars. Stainless steel bars and galvanized carbon steel bars and fabric are used to provide enhanced corrosion resistance.

The following types of steel are most commonly used for reinforcement: ASTM A615 Grade 40 and 60 and ASTM A706 Grade 60. Other types of steel that are used for reinforcement include Portuguese A400 NR and A500 NR (Cabrita et al. 1996), BS 4449 and BS 4461, quenched and self-tempered steel, and AISI 316 (stainless steel). The mechanical properties of the various types of reinforcing steels can be found in the literature (e.g., Tao et al. 2013, Holmes et al. 1982).

Many tests have been performed to characterize the mechanical properties of reinforcing steels at elevated temperatures. Most of these tests pertain to classical hot-rolled and cold-worked steels. In some European countries, such steels have been gradually abandoned and replaced by quenched and self-tempered steels (i.e., steels for which hot rolling is followed by a special thermal treatment). For these steels, very few tests have been conducted at elevated temperatures. Because of the lack of experimental data, the properties given for classical hot-rolled steels are also used for quenched and self-tempered steels, thus neglecting the effects of the additional cooling and tempering treatments on the mechanical characteristics at elevated temperatures.

For reinforced concrete beams and slabs exposed to fire, heating of tensile reinforcing bars near the exposed undersurface leads to thermal expansion and reductions in elastic modulus, both of which contribute to elongation of the bars. Elongation of tensile reinforcing bars results in increased deflections and allows tensile cracks in the concrete to open, which exposes the bars to accelerated heating. These effects can be exacerbated by reductions in yield strength at elevated temperature, which can result in plastic deformation associated with significantly increased elongation of the reinforcing bars.

Quenched-and-tempered (QT) steel reinforcement is primarily used for bridge structures in the United States. ASTM A514 is a common steel grade that is used for such reinforcement, but other proprietary steel grades are also used. Stress–strain data obtained from steady-state elevated-temperature tests for two different QT bar sizes are provided by Dotreppe (1997). Also, Seif et al. (2016a) demonstrates that the equations provided for retained yield strength and stress–strain data in Section 8.2.5.1 can be calibrated to QT reinforcing bars.

Eurocode 2 (CEN 2004) provides the mechanical properties of carbon steel reinforcement at elevated temperatures but not those of other types of reinforcement. This code treats hot-rolled reinforcement as hot-rolled carbon steel but introduces different models for cold-worked reinforcement. Figure 8-51 plots the stress–strain relationship of reinforcement steel based on work by Anderberg (1988).

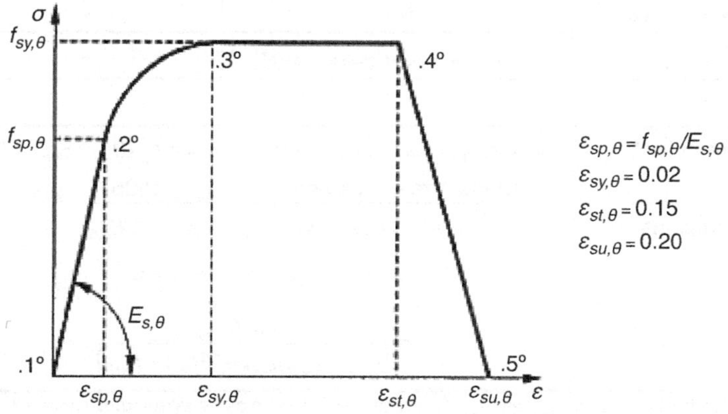

Figure 8-51. Stress–strain relationship of reinforcement steel at elevated temperatures.
Source: Eurocode 2 (CEN 2004).

Eurocode 2 (CEN 2004) divides reinforcement steel into two classes: Class N and Class X. Class N refers to *normal hardening* reinforcement. Although the choice of Class N is generally recommended, Class X may be used when there is experimental evidence for these values. Based on the limited amount of test results that are available, quenched and self-tempered reinforcing steel may be considered as Class X reinforcing steel. Most of the traditional hot-rolled and cold-worked steels fall in the Class N category. According to this code, Class N steels mostly comprise three steel classes that are used as bars and rods (i.e., Classes A, B, and C). These classes differ mostly in ductility and characteristic strength ratio. Class A is the lowest ductility category, Class B is most commonly used for reinforcing bars, and Class C is the highest ductility category. Expressions for stress–strain relationships at elevated temperatures are given in Figure 8-52 for the case when a Class A reinforcement is used.

Table 8-15 and Figure 8-53 present strength reduction factors for Class N reinforcement, including hot-rolled and cold-worked types. Note that Class N broadly covers all cases when any one of Classes A, B, or C steel reinforcement types is used. The reduction factors of proportional limit,

Range	Stress $\sigma(\theta)$	Tangent modulus
$\varepsilon_{sp,\theta}$	$\varepsilon E_{s,\theta}$	$E_{s,\theta}$
$\varepsilon_{sp,\theta} \leq \varepsilon \leq \varepsilon_{sy,\theta}$	$f_{sp,\theta} - c + (b/a)[a^2 - (\varepsilon_{sy,\theta} - \varepsilon)^2]^{0,5}$	$\dfrac{b(\varepsilon_{sy,\theta} - \varepsilon)}{a\left[a^2 - \left(\varepsilon - \varepsilon_{sy,\theta}\right)^2\right]^{0,5}}$
$\varepsilon_{sy,\theta} \leq \varepsilon \leq \varepsilon_{st,\theta}$	$f_{sy,\theta}$	0
$\varepsilon_{st,\theta} \leq \varepsilon \leq \varepsilon_{su,\theta}$	$f_{sy,\theta}[1-(\varepsilon - \varepsilon_{st,\theta})/(\varepsilon_{su,\theta} - \varepsilon_{st,\theta})]$	-
$\varepsilon = \varepsilon_{su,\theta}$	0,00	-
Parameter *)	$\varepsilon_{sp,\theta} = f_{sp,\theta} / E_{s,\theta}$ $\quad \varepsilon_{sy,\theta} = 0,02$ $\quad \varepsilon_{st,\theta} = 0,15$ $\quad \varepsilon_{su,\theta} = 0,20$ Class A reinforcement: $\varepsilon_{st,\theta} = 0,05$ $\quad \varepsilon_{su,\theta} = 0,10$	
Functions	$a^2 = (\varepsilon_{sy,\theta} - \varepsilon_{sp,\theta})(\varepsilon_{sy,\theta} - \varepsilon_{sp,\theta} + c/E_{s,\theta})$ $b^2 = c\,(\varepsilon_{sy,\theta} - \varepsilon_{sp,\theta})\,E_{s,\theta} + c^2$ $c = \dfrac{(f_{sy,\theta} - f_{sp,\theta})^2}{(\varepsilon_{sy,\theta} - \varepsilon_{sp,\theta})E_{s,\theta} - 2(f_{sy,\theta} - f_{sp,\theta})}$	

*) Values for the parameters $\varepsilon_{pt,\theta}$ and $\varepsilon_{pu,\theta}$ for prestressing steel may be taken from Table 3.3. Class A reinforcement is defined in Annex C of EN 1992-1-1.

Figure 8-52. Stress–strain relationships of Class A reinforcing steel at elevated temperatures.
Source: Eurocode 2 (CEN 2004).

Table 8-15. Reduction Factors for Stress–Strain Relationship of Class N Reinforcing Steel at Elevated Temperatures.

Steel Temperature θ[°C]	$f_{sy,\theta} / f_{yk}$		$f_{sp,\theta} / f_{yk}$		$E_{s,\theta} / E_s$	
	hot rolled	cold worked	hot rolled	cold worked	hot rolled	cold worked
1	2	3	4	5	6	7
20	1,00	1,00	1,00	1,00	1,00	1,00
100	1,00	1,00	1,00	0,96	1,00	1,00
200	1,00	1,00	0,81	0,92	0,90	0,87
300	1,00	1,00	0,61	0,81	0,80	0,72
400	1,00	0,94	0,42	0,63	0,70	0,56
500	0,78	0,67	0,36	0,44	0,60	0,40
600	0,47	0,40	0,18	0,26	0,31	0,24
700	0,23	0,12	0,07	0,08	0,13	0,08
800	0,11	0,11	0,05	0,06	0,09	0,06
900	0,06	0,08	0,04	0,05	0,07	0,05
1000	0,04	0,05	0,02	0,03	0,04	0,03
1100	0,02	0,03	0,01	0,02	0,02	0,02
1200	0,00	0,00	0,00	0,00	0,00	0,00

Source: Bailey (2017).

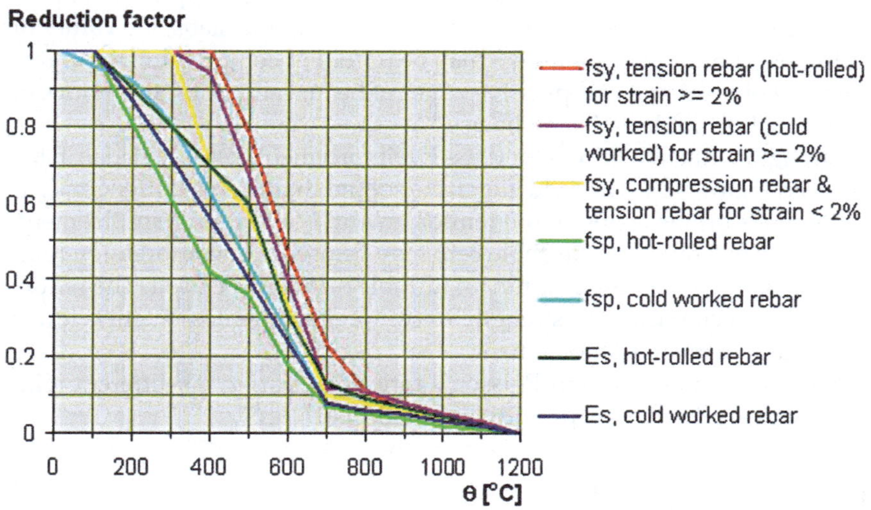

Figure 8-53. Reduction factors for stress–strain relationship of Class N reinforcing steel at elevated temperatures.
Source: Bailey (2017).

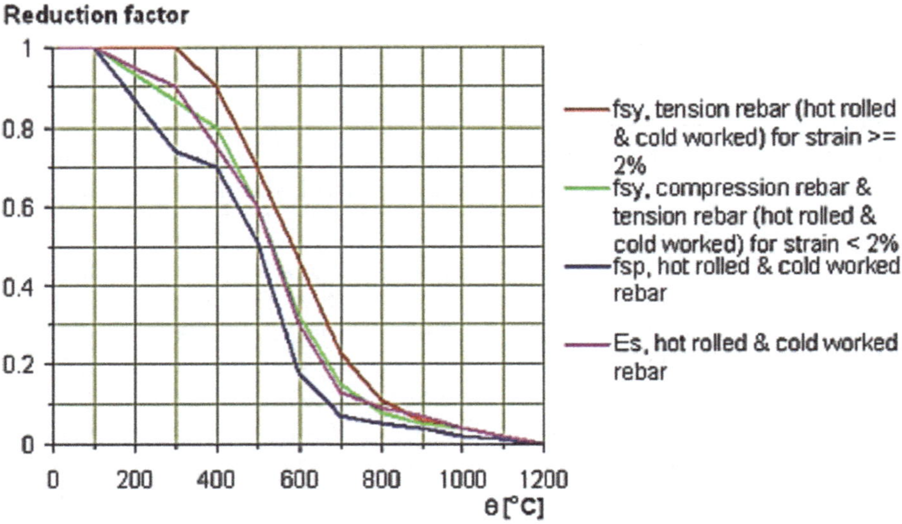

Figure 8-54. Reduction factors for stress–strain relationship of Class X reinforcing steel at elevated temperatures.
Source: Bailey (2017).

$f_{sp,\theta}/f_{yk}$, and the slope of the linear elastic range, $E_{s,\theta}/E_s$, are applicable to both tension and compression reinforcement. However, the reduction factor of the characteristic strength, $f_{sy,\theta}/f_{yk}$, depends on the application of the reinforcement. For tension reinforcement in beams and slabs, the values of Columns (2) and (3) in Table 8-15 may be used. For compression reinforcement in columns and compressive zones of beams and slabs, the values of Column (4) in Table 8-15 may be used.

Strength reduction factors for Class X reinforcement are provided in Table 8-16 and Figure 8-54. The reduction factor of the characteristic strength, $f_{sy,\theta}/f_{yk}$, depends on the action in the reinforcement. For tension reinforcement, the values of Column (2) in Table 8-16 may be used. For compression reinforcement, the values of Column (3) in Table 8-16 may be used. The reduction factor of characteristic strength for compression reinforcement is lower than that of tension reinforcement.

The elastic modulus and Poisson's ratio for a variety of different reinforcement steel grades is provided by Rothman (1988).

8.3.3 Prestressing Steel

Prestressing steel is made by cold-working high-carbon steel wires, which are then thermomechanically treated to improve relaxation properties and to create the prescribed mechanical properties. The cold-working process

Table 8-16. Reduction Factors for Stress–Strain Relationship of Class X Reinforcing Steel at Elevated Temperatures.

Steel Temperature θ[°C]	$f_{sy,\theta} / f_{yk}$ hot rolled and cold worked	$f_{sp,\theta} / f_{yk}$ hot rolled and cold worked	$E_{s,\theta} / E_s$ hot rolled and cold worked
20	1,00	1,00	1,00
100	1,00	1,00	1,00
200	1,00	0,87	0,95
300	1,00	0,74	0,90
400	0,90	0,70	0,75
500	0,70	0,51	0,60
600	0,47	0,18	0,31
700	0,23	0,07	0,13
800	0,11	0,05	0,09
900	0,06	0,04	0,07
1000	0,04	0,02	0,04
1100	0,02	0,01	0,02

Source: Bailey (2017).

increases the tensile strength of the wires owing to work hardening. The wires are heated and slowly cooled through a solid-state transformation called a *eutectoid reaction*. The fabrication process provides the strands with the required low-relaxation properties to ensure long-lasting performance in the prestressed concrete member. This forming process impacts the wire's mechanical properties at elevated temperatures, as well as its residual properties after cooling following a fire exposure.

8.3.3.1 Tensile Strength. The 0.2% offset yield strength and the ultimate tensile strength of prestressing steels decrease significantly at elevated temperatures (Anderberg 1983, Harmathy and Stanzak 1970, Holmes et al. 1982, ASCE 1992). Holmes et al. (1982) measured the elevated-temperature effects on prestressing steels that were manufactured to British Standard specifications. In this study, three types of prestressing steels were tested: 5 mm mill coil wire, 5 mm stabilized (low-relaxation) wire, and a 9.3 mm seven-wire strand. The ambient-temperature mechanical properties of these wires and strands are provided in Table 8-17. The yield (2% proof) and tensile strength of the prestressing wires and strands as a function of temperature are plotted in Figure 8-55.

8.3.3.2 Stress–Strain Response. The following equations provide high-temperature polynomial relationships for yield stress and tensile

Table 8-17. Ambient-Temperature Properties of Prestressing Steels.

Type of Steel	Yield (2% proof) stress (MPa)	Tensile strength (MPa)	Elastic modulus (GPA)
5 mm mill coil	160.6	369	203
5 mm stabilized wire	146.5	335	204
9.3 mm seven-wire strand	186.4	104	200

Source: Holmes (1982).

strength of prestressing strands conforming to ASTM A416 (Shakya and Kodur 2016):

$$\frac{\sigma_{yT}}{\sigma_y} = 5.925E^{-12}T^4 - 5.898E^{-09}T^3 + 1.455E^{-07}T^2 - 5.978E^{-04}T + 1.020 \quad \text{(8-15)}$$

$$\frac{\sigma_{uT}}{\sigma_u} = 1.332E^{-12}T^4 + 3.347E^{-09}T^3 - 5.778E^{-06}T^2 + 6.346E^{-04}T + 0.9894 \quad \text{(8-16)}$$

where
σ_y, σ_{yT} = yield stress (MPa) at ambient and temperature T, respectively
σ_u, σ_{uT} = ultimate strength (MPa) at ambient and temperature T, respectively

8.3.3.3 Elastic Modulus. Holmes et al. (1982) provided reduction factors for the elastic moduli of mill coil wire, stabilized (low-relaxation) wire, and a seven-wire strand at elevated temperatures. These reductions were normalized to the ambient-temperature values provided in Table 8-17. Similarly, Figure 8-56 provides elastic modulus as a function of temperature that is normalized to a stress-relieved wire with a modulus of 214 GPa at ambient temperature (Ruge and Winkelmann 1978, Anderberg 1983). A regression analysis of this data is presented as the following second-degree polynomial:

$$\frac{E_T}{E_{20^oC}} = -2 \times 10^{-6}T^2 + 0.2 \times 10^{-6}T + 0.987 \quad \text{(8-17)}$$

8.3.3.4 Creep. There is a gradual loss of prestress that occurs up to approximately 315°C, which is followed by a more drastic decrease above this temperature. Sources in the literature provide models to predict the transient creep–relaxation of prestressing tendons at elevated temperature (Abrams and Cruz 1961, MacLean 2007). Wei et al. (2016) and Gales et al.

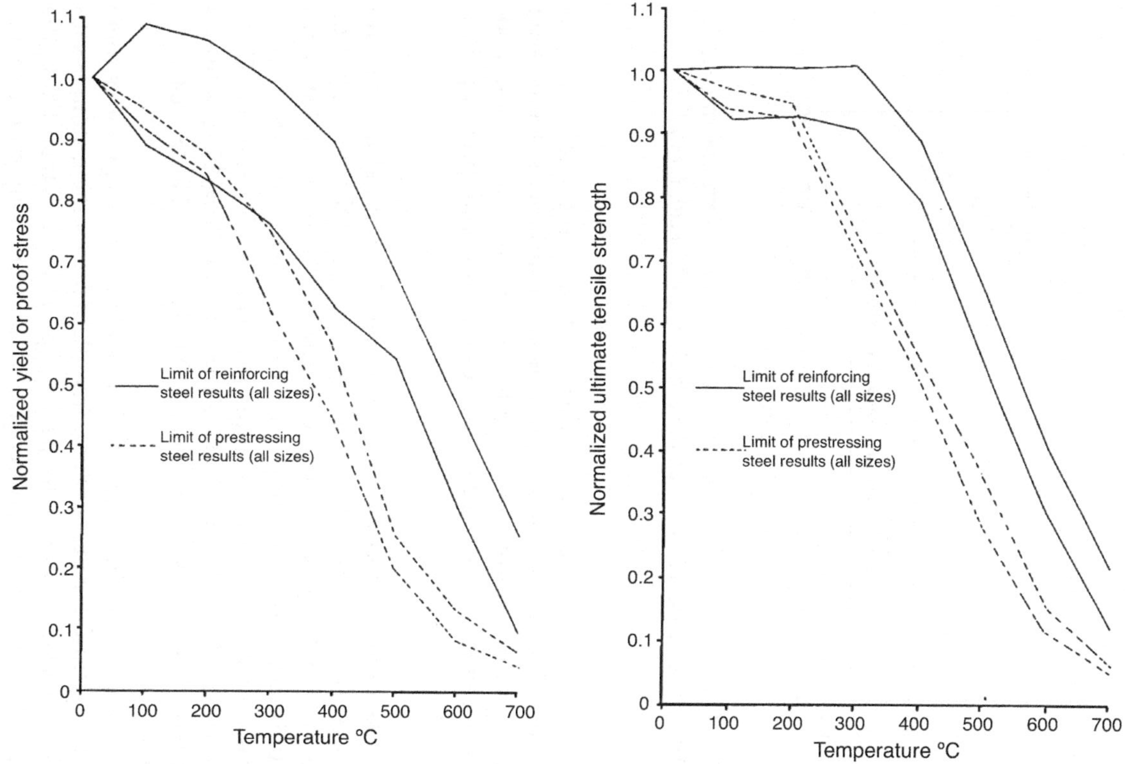

Figure 8-55. Normalized yield stress and tensile strength of prestressing steel at elevated temperatures. Source: Holmes et al. (1982).

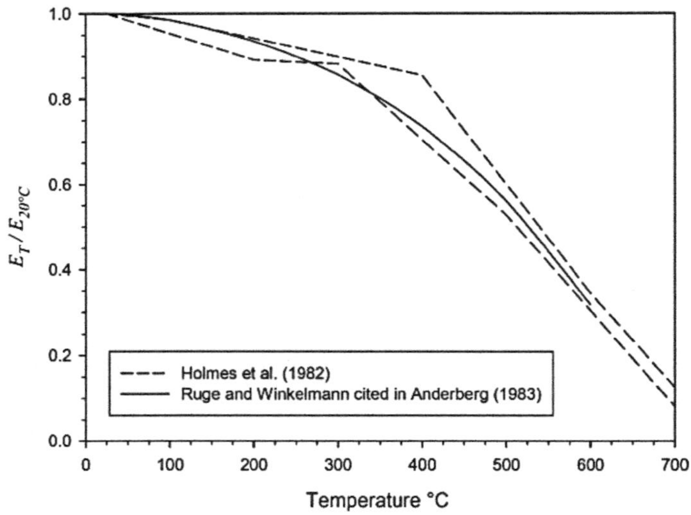

Figure 8-56. Effects of temperature on the elastic modulus of prestressing steel. Sources: Anderberg (1983), Holmes et al. (1982), and MacLean (2007).

(2016) modified these models to include the tertiary creep stage characterized by increasing strain rate until rupture, resulting in an overall improved prediction of prestress loss owing to high-temperature exposure. These models should be carefully evaluated by the designer for appropriate application prior to implementation.

8.3.3.5 Residual Properties. The residual elastic modulus of prestressing steel wires has been observed not to change if the steel is heated below 720°C (Fan 2004). However, the residual 0.2% offset yield strength and the ultimate tensile strengths of various prestressing steels have been shown to decrease significantly between 400°C and 600°C, as shown in Figure 8-57 (Guyon 1953; Abrams and Cruz 1961; Abrams and Erlin 1967; Holmes 1982; Neves et al. 1996; Fan 2004; Hertz 2004; MacLean 2007).

The Precast Concrete Institute (PCI) manual (2011) also provides some guidance for quantifying the residual strength of prestressing strands subjected to elevated temperatures: for peak temperatures less than 400°C, the strand retains 100% of its initial strength, but for peak temperatures equal to 482°C and 600°C, it retains 70% and 50% of its initial strength, respectively. These values are similar to but not the same as those reported by Abrams and Erlin (1967). Hertz (2004) developed a closed-form solution that predicts the residual strength of various building materials after being subjected to elevated temperature as follows:

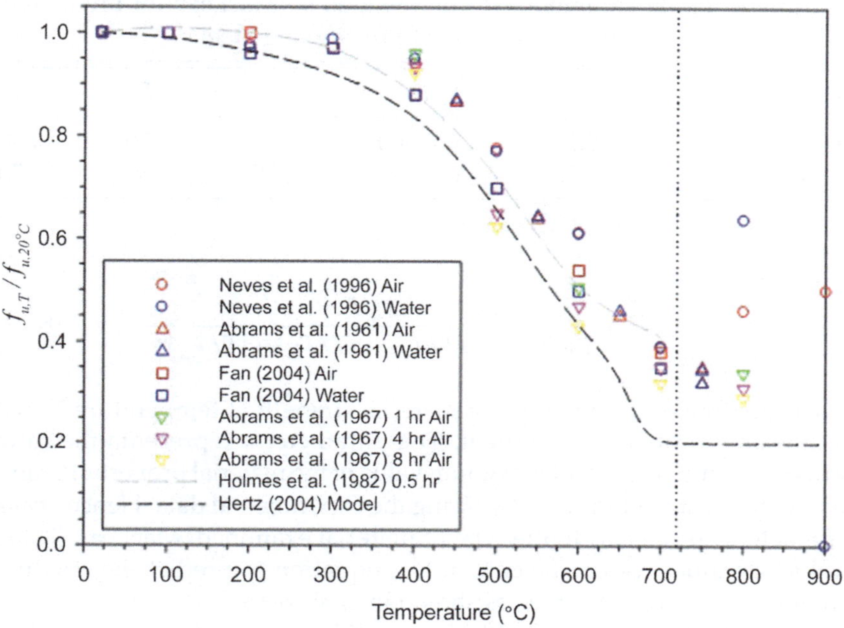

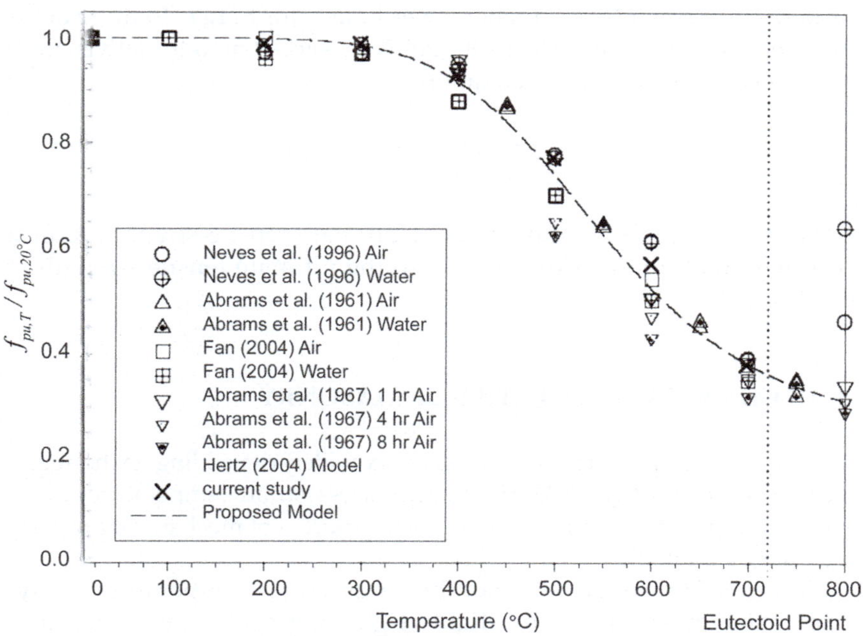

Figure 8-57. Residual yield and tensile strength of prestressing steels. Source: MacLean (2007).

Table 8-18. Hertz Parameters Recommended by MacLean for Predicting the Residual Stress of Prestressing Steel.

Residual stress	k	T_1	T_2	T_8	T_{64}
Yield	0.20	100 000	750	550	650
Ultimate	0.20	100 000	950	550	650

Sources: Hertz (2004) and MacLean (2007).

$$\xi(T) = k + \frac{1-k}{1+(T/T_1)+(T/T_2)^2+(T/T_8)^8+(T/T_{64})^{64}} \tag{8-18}$$

The parameter ξ is the ratio of material strength at temperature T (°C) to the same property at ambient. The parameter k represents the ratio between the minimum and maximum value of the material property. Parameters T_1 to T_{64} are determined by fitting the experimental data. Hence, these parameters are specific to the type of material examined. MacLean (2007) provided values for coefficients in this equation to predict the residual strength of prestressing steel, as shown in Table 8-18.

The Hertz model using the MacLean (2007) parameters is conservative, but perhaps it provides a suitable lower bound for SFED. To more accurately represent the data, MacLean (2007) presented an alternative curve represented by the following equation:

$$f_u(T) = 0.25 + \frac{0.75}{1+(T/550)^{6.5}} \tag{8-19}$$

The parameter $f_u(T)$ is the ratio of ultimate tensile strength after a given temperature exposure, T (°C), to the ultimate tensile strength at ambient.

8.4 MECHANICAL PROPERTIES OF MASONRY

The mechanical properties of masonry vary widely depending on the aggregate type. Eurocode 6 (CEN 2005) provides stress–strain–temperature curves for certain types of masonry. Other industry sources of mechanical property data are not available.

Figures 8-58 through 8-60 plot the stress–strain–temperature behavior of calcium silicate units, lightweight aggregate units (pumice), and autoclaved aerated concrete units, respectively (Eurocode 6, CEN 2005).

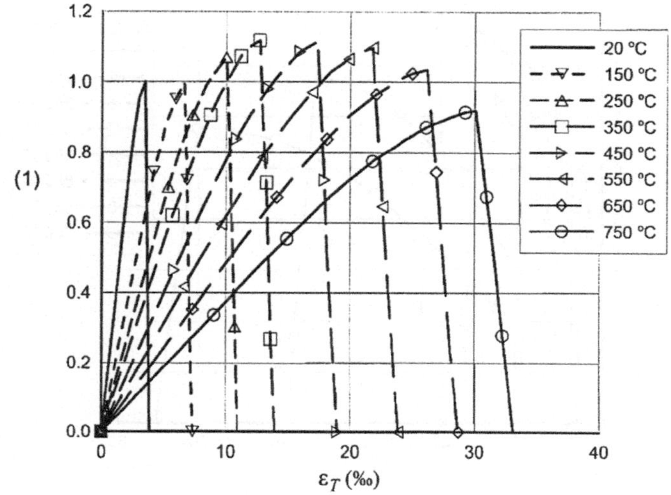

Figure 8-58. Stress–strain behavior of calcium silicate units (solid) at elevated temperatures with a normalized compressive strength range of 12 to 20 N/mm^2 and with a gross dry density range of 1,600 to 2,000 kg/m^3.
Source: Eurocode 6 (CEN 2005).

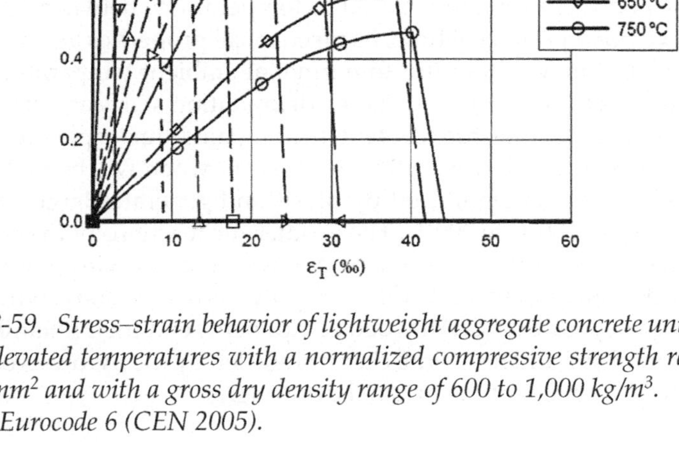

Figure 8-59. Stress–strain behavior of lightweight aggregate concrete units (pumice) at elevated temperatures with a normalized compressive strength range of 4 to 6 N/mm^2 and with a gross dry density range of 600 to 1,000 kg/m^3.
Source: Eurocode 6 (CEN 2005).

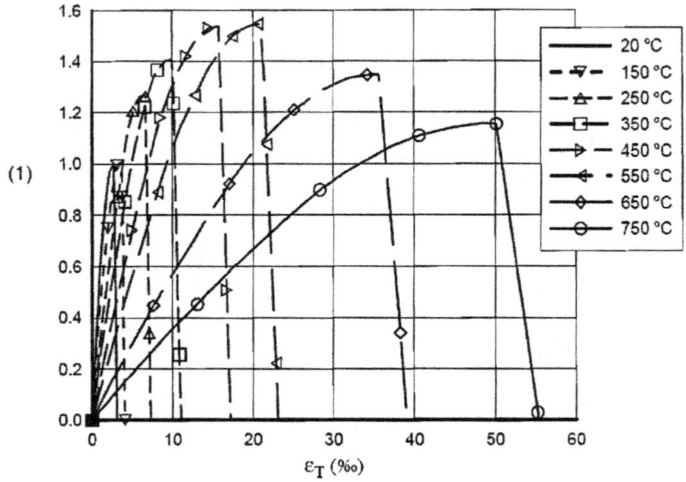

Figure 8-60. Stress–strain behavior of autoclaved aerated concrete units at elevated temperatures with a normalized compressive strength range of 4 to 6 N/mm² and with a gross dry density range of 400 to 600 kg/m³.
Source: Eurocode 6 (CEN 2005).

8.5 MECHANICAL PROPERTIES OF TIMBER

Tree growth is subject to many natural influences (i.e., weather, soil conditions, and growing space), so the mechanical properties of timber vary considerably. Notably, the mechanical properties parallel and perpendicular to the grain orientation differ. All mechanical properties used for timber design are derived from extensive sampling and analysis procedures, and these properties are represented as either minimum strength or average stiffness.

When wood members are exposed to temperatures between 38°C and 66°C for extended periods of time, its mechanical properties are temporarily reduced, but they will return to normal at ambient temperatures. For temperatures between 66°C and 100°C, degradation starts and its magnitude depends on the moisture content, heating medium, exposure period, and the wood species. Between 100°C and 200°C, wood will begin to pyrolyze. Wood undergoes significant pyrolysis and generates large amounts of CO gases from 200°C to 300°C. The exothermic reactions of the char and volatiles with surrounding oxygen are manifested as glowing combustion in this temperature regime. Significant depolymerization starts when temperatures go beyond 300°C. All wood components end their volatile emissions at around 450°C. The overall char residue left by the pyrolysis of wood is typically 25% or less of the original dry weight.

Because the strength and stiffness of the char layer are nearly zero, the properties of wood from room temperature to 300°C are of most interest to SFED.

8.5.1 Moisture Effects

The moisture content is an important parameter that affects the performance of wood when exposed to fire. However, the moisture content can be sensitive to the test method and the specimen size. Moisture migrates into the interior of the wood members during fire exposure, which causes moisture gradients that slow the progression of the char front in larger wood cross sections.

Wood becomes flexible at certain combinations of temperature and moisture content. For instance, wood can be bent to large angles using steam heating. Wood softening may occur when it is heated in a fire, but such softening lasts just a short period of time. The effect of softening is also highly dependent on the cross-section size. For a large cross-section wood member, steam softening only occurs in the thin outer layer and does not cause significant property changes to the member. However, for a small cross-section wood member, a large proportion of the cross section may experience steam softening, and thus it may become flexible in compression or bending. Because steam softening occurs for a short period of time at elevated temperatures, if the member can resist the applied loads, the mechanical properties of the wood can recover as the moisture is driven away.

8.5.2 Parallel-to-Grain Properties

The mechanical properties of wood generally decrease when heated and increase when cooled. There are both reversible effects and irreversible effects at elevated temperatures depending on the temperature magnitudes and the exposure duration. When wood is exposed to an immediate temperature change below 100°C and tested, the effect is essentially reversible. Accordingly, the mechanical properties will return to the value at ambient temperature if the elevated-temperature exposure is rapid.

Figure 8-61 plots the immediate effect of elevated temperatures on the elastic modulus, compression strength, and bending strength of wood (FPL 2010). The properties for dry wood (i.e., wood with less than 20% moisture content) at approximately 12% moisture content are provided. It should be noted that most handheld moisture meters do not record below 6% moisture content, in which case an oven dry test would be necessary for accurate determination of moisture content below 6%. The effect of temperature on the elastic modulus parallel to the grain is roughly linear up to 200°C, as shown in Figure 8-61(a). For compression

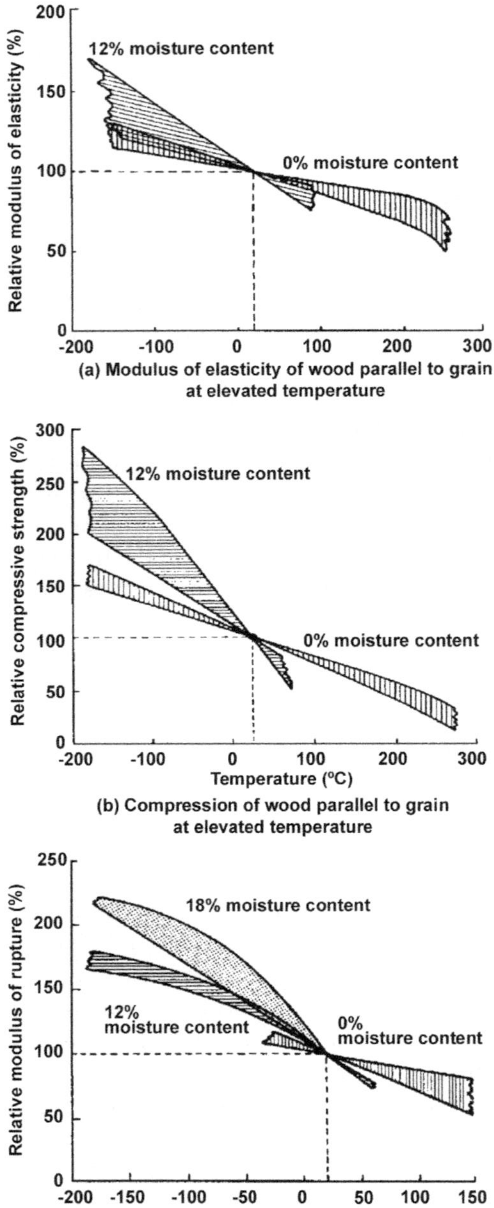

Figure 8-61. Reversible properties of wood.
Source: FPL (2010).

strength, moist wood is only tested up to 60°C, as shown in Figure 8-61(b). This is owing to the steam-softening behavior, as mentioned in a previous section. The steam-softening behavior also affects the bending behavior of wood at elevated temperatures. When the environmental conditions become suitable for steam-softening behavior, the compression zone can increase significantly, which causes a shift in the neutral axis, resulting in large deformations. Figure 8-61(c) represents the bending strength at elevated temperature.

There are also irreversible effects to wood members at elevated temperatures. These permanent effects result in loss of weight and strength, and they heavily depend on the moisture content, heating medium, temperature, and exposure duration and also on the species and size of the member (FPL 2010). The temperature and heating duration have permanent effects on the modulus of rupture caused by heating in steam and water as shown in Figure 8-62.

8.5.3 Perpendicular-to-Grain Properties

There is a lack of any significant research on the perpendicular-to-grain mechanical properties of wood at elevated temperatures. In cases where the perpendicular-to-grain properties of wood members are important, the designer should proceed with caution and perhaps seek validated test data to properly conduct an SFED.

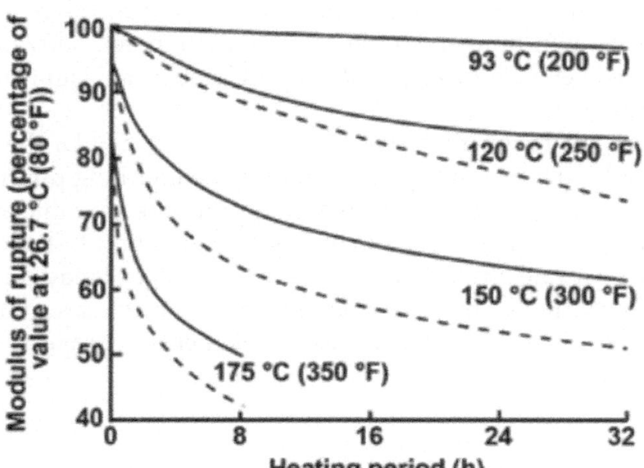

Figure 8-62. Permanent effect of heating in water (solid line) and steam (dashed line) on modulus of rupture.
Source: FPL (2010).

Table 8-19. Effective Char Rates Based on Standard Fire Exposure (for $\beta n = 1.5$ in./h).

Fire Exposed Time (h)	Effective Char Rate, β_{eff} (in./h)	Effective Char Layer Thickness, a_{eff} (in.)
1	1.80	1.8
1.5	1.67	2.5
2	1.58	3.2

Source: Derived from AWC (2015).

8.5.4 Char Rates

Knowledge of char rates is important for determining the mechanical properties of wood at elevated temperatures. Char rates are also discussed in Section 5.7. The effective char rate for wood members exposed to the standard fire (per SFRD) can be estimated as follows (AWC 2015):

$$\beta_{eff} = \frac{1.2\,\beta_n}{t^{0.187}} \tag{8-20}$$

where

β_{eff} = effective char rate (in./h), adjusted for exposure time, t

β_n = nominal char rate (in./h), linear char rate based on 1-h exposure

t = exposure time (h)

Considering these relations, the effective char rate and char layer thickness for solid-sawn lumber, structural glued-laminated softwood members, laminated veneer lumber, parallel strand lumber, and laminated strand lumber are given in Table 8-19.

For structural design fires that are applicable to SFED, the designer should consider the relations presented for benchmarking purposes only. Moreover, the designer should derive an appropriate char rate applicable to structural design fires, which may require validation testing. Once an appropriate char rate has been established, the designer may consider the remaining uncharred section for structural analysis purposes, granting that the possible loss of any adhesion between elements is properly analyzed and validated.

REFERENCES

Abrams, M. S., and C. R. Cruz. 1961. "Behavior at high temperature of steel strand for prestressed concrete." *J. PCA. Res. Dev. Lab.* 3(3), 8–19.

Abrams, M., and B. Erlin. 1967. "Estimating post-fire strength and exposure temperature of prestressing steel by metallographic method." *J. PCA Res. Dev. Lab.* 9(3), 23–33.

ACI (American Concrete Institute). 2014. *Code requirements for determining fire resistance of concrete and masonry construction assemblies*. Standard ACI 216.1-07/TMS-0216-07. Farmington Hills, MI: ACI.

AISC (American Institute of Steel Construction). 2016. *Specification for structural steel buildings*. Standard ANSI/AISC 360-16. Chicago: AISC.

AISI (American Iron and Steel Institute). 2016. *North American specification for the design of cold-formed steel structural members*. S100. Washington, DC: AISI.

Anderberg, Y. 1976. "Fire-exposed hyperstatic concrete structures—An experimental and theoretical study." *Bull. Div. Struct. Mech. Concrete Const. 55*. Lund, Sweden: Lund Institute of Technology.

Anderberg, Y. 1983. "Predicted fire behaviour of steels and concrete structures." *LUTVDG/TVBB—3011—SE, Vol. 3011*, Division of Building Fire Safety and Technology. Lund, Sweden: Lund Institute of Technology.

Anderberg, Y. 1988. "Modelling steel behaviour." *Fire Saf. J.* 13(1), 17–26.

AS (Australian Standard). 1998. *Steel structures*. AS 4100. Sydney, Australia: AS.

ASCE. 1992. "Structural fire protection." *Tech. Rep. No. 78*, T. T. Lie (ed.). New York: ASCE.

Assefpour-Dezfuly, M., B. Hugaas, and A. Brownrigg. 1990. "Fire resistant high-strength low-alloy steels." *Mater. Sci. Technol.* 6(12), 1210–1214.

ASTM A1077/A1077M. 2012. *Standard specification for structural steel with improved yield strength at high temperature for use in buildings*. West Conshohoken, PA: ASTM International.

AWC (American Wood Council). 2015. "Calculating the fire resistance of exposed wood members." *TR 10*. Leesburg, VA: AWC.

Bailey, C. 2007. "Stainless steel introduction." Accessed May 9, 2018. http://www.mace.manchester.ac.uk/project/research/structures/strucfire/materialInFire/Steel/StainlessSteel/default.htm?p=4|#4.

Bailey, C. 2016. "Fire-resistant steel mechanical properties." Accessed May 9, 2018. http://www.mace.manchester.ac.uk/project/research/structures/strucfire/materialInFire/Steel/FireResistantSteel/mechanicalProperties.htm?p=8|#8.

Bailey, C. 2017. "Reinforcing steel mechanical properties." One Stop Shop in Structural Fire Engineering. Manchester, UK: Univ. of Manchester.

Batista-Abreu, J. 2015. "Fire performance of cold-formed steel walls." Ph.D. dissertation, Baltimore: Johns Hopkins Univ., Dept. of Civil Engineering.

Behnood, A., and M. Ghandehari. 2009. "Comparison of compressive and splitting tensile strength of high-strength concrete with and without polypropylene fibers heated to high temperatures." *Fire Saf. J.* 44(8), 1015–1022.

British Steel (1983). "The behaviour of structural steels manufactured by BSC under stress controlled anisothermal creep conditions." *Rep. SH/RS/3664/4/83/B.*

BS (British Standard). 1998. *Structural use of steelwork in building—Part 8: Code of practice for fire resistant design.* British Standard BS 5950-8:1990. London: British Standards Institution.

Cabrita Neves, I. C., J. P. C. Rodrigues, and A. de Pádua Loureiro. 1996. "Mechanical properties of reinforcing and prestressing steels after heating." *J. Mater. Civ. Eng.* 8(4), 189–194.

Calado, W. R., O. J. dos Santos, C. S. B. Castro, R. N. Barbosa, and B. M. Gonzalez. 2008. "Effect of finishing rolling temperature on fire resistance and dynamic strain aging behavior of a structural steel." *J. Mater. Sci.* 43(17), 6005–6011.

Carette, G. G., K. E. Painter, and V. M. Malhotra. 1982. "Sustained high-temperature effect on concretes made with normal Portland cement, normal Portland cement and slag, or normal Portland cement and fly ash." *Concr. Int.* 4(7), 16.

Castillo, C., and A. J. Durrani. 1990. "Effect of transient high temperature on high-strength concrete." *ACI Mater. J.* 87(1), 47–53.

CEN (European Committee for Standardization). 2004. *Eurocode 2—Design of concrete structures. Part 1–2: General rules—Structural fire design.* EN 1992-1-2. Brussels: CEN.

CEN. 2005. *Eurocode 6—Design of masonry structures. Part 1–2: General rules—Structural fire design.* EN 1996-1-2. Brussels: CEN.

CEN. 2006. *Eurocode 3—Design of steel structures. Part 1–2: General rules—Structural fire design.* EN 1993-1-2. Brussels: CEN.

Chen, W., and J. Ye. 2012. "Mechanical properties of G550 cold-formed steel under transient and steady state conditions." *J. Constr. Steel Res.* 73, 1–11.

Chen, J., and B. Young. 2006. "Stress–strain curves for stainless steel at elevated temperatures." *Eng. Struct.* 28(2), 229–239.

Chen, J., B. Young, and B. Uy. 2006. "Behavior of high strength structural steel at elevated temperatures." *J. Struct. Eng.* 132(12), 1948–1954.

Chiew, S. P., M. S. Zhao, and C. K. Lee. 2014. "Mechanical properties of heat-treated high strength steel under fire/post-fire conditions." *J. Constr. Steel Res.* 98, 12–19.

Chijiwa, R., H. Tamehiro, Y. Yosida, K. Funato, and R. Uemori. 1993. *Bull. Jpn. Inst. Metals* 32, 432–434.

Chijiiwa, R., H. Tamehiro, A. Yoshie, K. Funato, Y. Yoshida, Y. Horii, et al. 1995. "Development and practical application of fire-resistant steel for buildings." *Proc., 1995 Conf. on HSLA Steels*, China Science and Technology Press, Beijing.

Choe, L., C. Zhang, W. E. Luecke, J. L. Gross, and A. H. Varma. 2017. "Influence of material models on predicting the fire behavior of steel columns." *Fire Technol.* 53(1), 375–400.

Cooke, G. 1988. "An introduction to mechanical properties of structural steel at elevated temperatures." *Fire Saf. J.* 13(1), 1–74.

Creusot-Loire. 1976. "Recherce et mise au point d'acier moyennment allies a resistance au feu renforcee: Etude theorique des caracteristiques optimals." *75-61542-06-223-75-01, Raport Techniques No. 1*, Creusot-Loire, Usine du Creusot (in French).

Creusot-Loire. 1977. "Recherche et Mise au Point d'Acier Moyennment Allies a Resistance au Feu Renforcee, Determination des Caracteristiques a Chaud des Aciers." *75-61542-06-223-75-01, Raport Techniques No. 2*, Creusot-Loire, Usine du Creusot (in French).

Dotreppe, J. 1997. "Mechanical properties of quenched and self-tempered reinforcing steel at elevated temperatures compared with recommendations of Eurocode 2, Part 1–2." *Mater. Struct.* 30(7), 430–438.

Dwaikat, M. 2009. "Flexural response of reinforced concrete beams exposed to fire." Ph.D. thesis, East Lansing, MI: Michigan State Univ., Dept. of Civil Engineering.

Dwaikat, M. B., and V. K. R. Kodur. 2009. "Response of restrained concrete beams under design fire exposure." *J. Struct. Eng.* 135(11), 1408–1417.

ECCS (European Convention for Constructional Steelwork, Technical Committee 3). 2001. "Model code on fire engineering." *Document No. 111*. Brussels: ECCS.

Fan, J. 2004. "Experimental study on material properties of prestressed steel strand post high temperatures." *Nanjing Ligong Daxue Xuebao/ J. Nanjing Univ. Sci. Technol.* 28(2).

Felicetti, R., P. G. Gambarova, G. Rosati, F. Corsi, and G. Giannuzzi. 1996. "Residual mechanical properties of high-strength concretes subjected to high-temperature cycles." *Proc., Int. Symp. on Utilization of High-Strength/ High-Performance Concrete*, Paris. Accessed June 5, 2018. https://www.hindawi.com/journals/isrn/2014/468510/ref/.

FPL (Forest Products Laboratory). 2010. "Wood handbook: Wood as an engineering material." *Forest Products Laboratory General Tech. Rep. FPL-GTR-190*. Madison, WI: USDA Forest Service.

Fujimoto, M., F. Furumura, T. Ave, and Y. Shinohara. 1980. "Primary creep of structural steel (SS 41) at high temperatures." *Trans. Arch. Inst. Jpn.* 296, 145–157.

Fujimoto, M., F. Furumura, and T. Ave. 1981. "Primary creep of structural steel (SM 50A) at high temperatures." *Trans. Arch. Inst. Jpn.* 306, 148–156.

Fujimoto, M., F. Furumura, and T. Ave. 1982. "Primary creep of structural steel (SM 58Q) at high temperatures." *Trans. Arch. Inst. Jpn.* 319, 147–155.

Fujino H., K. Hitomi, J. Hashimoto, and S. Umezawa. 1993. "Fire-resistant steel for building structures." Accessed June 5, 2018. http://www.company

-histories.com/KAWASAKI-STEEL-CORPORATION-Company-History.html.

Gales, J., L. Robertson, and L. Bisby. 2016. "Creep of prestressing steels in fire." *Fire Mater.* 40(7), 875–895.

Gonzalez, F., and J. Lange. 2009. "Behavior of high-strength grade 10.9 bolts under fire conditions." *Proc., Application of Structural Fire Design*, Prague, Czech Republic. Accessed June 5, 2018. http://fire.fsv.cvut.cz/ASFE09/Proceedings/0_List_of_contents.pdf.

Gosain, N., R. Drexler, and D. Choudhuri. 2008. "Evaluation and repair of fire-damaged buildings, *Struct. Mag.* Accessed May 29, 2018. http://www.structuremag.org/wp-content/uploads/2014/08/C-STRForensics-Fire-Gosain-Sept081.pdf.

Guyon, Y. 1953. *Prestressed concrete*. London: Contractors Record and Municipal Engineering.

Hanus, F., G. Zilli, and J. Franssen. 2011. "Behaviour of Grade 8.8 bolts under natural fire conditions—Tests and model." *J. Constr. Steel Res.* 67(8), 1292–1298.

Harmathy, T. 1967. "A comprehensive creep model." *J. Basic Eng.* 89(3), 496–502.

Harmathy, T., and W. Stanzak. 1970. "Elevated-temperature tensile and creep properties of some structural and prestressing steels. In *Fire test performance, ASTM STP 464*. West Conshohocken, PA: ASTM.

Harvey, P. D. 1982. *Engineering properties of steel*. Metals Park, OH: ASM.

Hertz, K. 2004. "Reinforcement data for fire safety design." *Mag. Concr. Res.* 56(8), 453–459.

Holmes, M., R. Anchor, G. Cook, and R. Crook. 1982. "The effects of elevated temperatures on the strength properties of reinforcing and prestressing steels." *Struct. Eng.* 60(1), 7–13.

Huang, L. (2017). Research on mechanical properties of domestic high strength structural steel at elevated temperatures, M.Sc. thesis, Shanghai: Tongji Univ., College of Civil Engineering.

Kand, S. 2011. "Effect of temperature on thermal and mechanical properties of high strength steel A325 and A490 bolts." M.Sc. thesis, East Lansing, MI: Michigan State Univ., Dept. of Civil Engineering.

Kankanamge, D. 2011. "Mechanical properties of cold-formed steels at elevated temperatures." *Thin-Walled Struct.* 49(1), 26–44.

Kawasaki Steel. 1993. *Tech. Rep. No. 29.* Accessed June 5, 2018. http://www.jfe-steel.co.jp/archives/en/ksc_giho/no.29/mokuji.htm.

Kelly, F., and W. Sha. 1999. "A comparison of the mechanical properties of fire-resistant and S275 structural steels." *J. Constr. Steel Res.* 50(3), 223–233.

Khaliq, W. 2012. "Performance characterization of high-performance concretes under fire conditions." Ph.D. thesis, East Lansing, MI: Michigan State Univ., Dept. of Civil Engineering.

Khoury, A., G. Grainger, and E. Sullivan. 1985. "Strain of concrete during first heating to 600°C." *Mag. Concr. Res.* 37(133), 195–215.

Kirby, B. 1995. "The behavior of high-strength grade 8.8 bolts in fire." *J. Constr. Steel Res.* 33(1-2), 3–38.

Kirby, B., and R. Preston. 1988. "High-temperature properties of hot-rolled, structural steels for use in fire engineering design studies." *Fire Saf. J.* 13(1), 27–37.

Klippstein, K. H. 1980. *Strength of cold-formed studs exposed to fire*. Washington, DC: AISC.

Knight, D., D. Skinner, and D. Lay. 1971. "Prediction of isothermal creep." *Report MRL 18/2*, Melbourne, Australia: BHP Melbourne Research Laboratories, Broken Hill Proprietary.

Kodur, V. 2014. "Properties of concrete at elevated temperatures." *ISRN Civ. Eng.* 468510. Accessed May 30, 2018. https://doi.org/10.1155/2014/468510.

Kodur, V. K. R. 2000. "Spalling in high-strength concrete exposed to fire—concerns, causes, critical parameters and cures." *Proc., ASCE Structures Congress: Advanced Technology in Structural Engineering*. Reston, VA: ASCE.

Kodur, V. K. R., S. Kand, and W. Khaliq. 2012. "Effect of temperature on thermal and mechanical properties of steel bolts." *J. Mater. Civ. Eng.* 24(6), 765–774.

Kodur, V. K. R., and E. M. Aziz. 2015. "Effect of temperature on creep in ASTM A572 high-strength low-alloy steels." *Mater. Struct.* 48(6), 1669–1677.

Lawson, R., and G. Newman. 1990. *Fire resistant design of steel structures—A handbook to BS 5950: Part 8, SCI Publication 080*. Berkshire, UK: Steel Construction Institute.

Lawson, R., B. Burgan, and G. Newman. 1993. *Building design using cold formed steel sections: Fire protection, SCI Publication P129*. Berkshire, UK: Steel Construction Institute.

Lee, J. 2012. "Elevated temperature properties of steel for structural fire engineering analysis." Ph.D. dissertation, Austin, TX: Univ. of Texas, Dept. of Civil, Architectural, and Environmental Engineering.

Lee, J., M. Mahendran, and P. Mäkeläinen. 2003. "Prediction of mechanical properties of light gauge steels at elevated temperatures." *J. Constr. Steel Res.* 59(12), 1517–1532.

Lee, J., A. Morovat, M. Engelhardt, and E. Taleff. 2012a. "Creep behavior of ASTM A992 steel at elevated temperatures." *Proc., Structures in Fire (SiF). Zurich, June 6–8*. Structures in Fire.

Lee, J., M. Engelhardt, and E. Taleff. 2012b. "Mechanical properties of ASTM A992 steel after fire." *AISC Eng. J.* 49(1), 33–44.

Lee, J., A. Morovat, G. Hu, M. Engelhardt, and E. Taleff. 2013. "Experimental investigation of mechanical properties of ASTM A992 steel at elevated temperatures." *AISC Eng. J.* 50(4), 249–272.

Lou, G., S. Yu, and R. Wang. 2010. "Experimental study of mechanical properties of high-strength bolts after fire." *Proc., 6th Int. Conf. on Structures in Fire*. Lancaster, PA: DEStech Publications, Inc.

Luecke, W. 2005. "Federal building and fire safety investigation of the World Trade Center disaster—Mechanical properties of structural steels." *Rep. NIST NCSTAR 1-3D*. Gaithersburg, MD: NIST.

Luecke, W., S. W. Banovic, and J. D. McColskey. 2011. "High-temperature tensile constitutive data and models for structural steels in fire." *NIST Tech. Note 1714*. Gaithersburg, MD: NIST.

MacLean, K. 2007. "Post-fire assessment of unbonded post-tensioned concrete slabs: Strand deterioration and prestress loss." M.S. thesis, Ontario, Canada: Queen's Univ. of Kingston, Dept. of Civil Engineering.

Mäkeläinen, P., and K. Miller. 1983. *Mechanical properties of cold-formed galvanized sheet steel Z32 at elevated temperatures*. Helsinki, Finland: Helsinki Univ. of Technology.

Matar, M. 2014. "Primary creep in ASTM A325 bolts under simulated fire loading." M.Sc. thesis, Milwaukee, WI: Univ. of Wisconsin. Accessed June 5, 2018. https://www.scribd.com/document/380454829/Primary-Creep-in-ASTM-A325-Bolts-Under-Simulated-Fire-Loading-pdf.

Morovat, M. A., J. Lee, M. D. Engelhardt, E. M. Taleff, T. A. Helwig, and V. A. Segrest. 2012. "Creep properties of ASTM A992 steel at elevated temperatures." *Adv. Mater. Res.* 446, 786–792.

Morovat, A. 2014. "Creep buckling behavior of steel columns subjected to fire." Ph.D. dissertation, Austin, TX: Univ. of Texas, Dept. of Civil Engineering.

Neves, I. C., J. P. C. Rodrigues, and A. de Pádua Loureiro. 1996. "Mechanical properties of reinforcing and prestressing steels after heating." *J. Mater. Civ. Eng.* 8(4), 189.

Outinen, J., O. Kaitila, and P. Mäkeläinen. 2001. "High-temperature testing of structural steel and modelling of structures at fire temperatures." *Res. Rep., Laboratory of Steel Structures Publications 23*. Helsinki, Finland: Helsinki Univ. of Technology.

Outinen, J., and P. Mäkeläinen. 2007. "Mechanical properties of structural steels at high temperatures and after cooling down." *Res. Rep., Laboratory of Steel Structures Publications 32*. Helsinki, Finland: Helsinki Univ. of Technology.

PCI (Precast Concrete Institute). 2011. *Design for fire resistance of precast/prestressed concrete*, 3rd Ed. Chicago: PCI.

Phan, L. 1996. "Fire performance of high-strength concrete: A report of the state-of-the-art." *Tech. Rep.* Gaithersburg, MD: NIST.

Qiang, X., F. Bijlaard, and H. Kolstein. 2013. "Post-fire performance of very high strength steel S960." *J. Constr. Steel Res.* 80, 235–242.

Qiang, X., X. Jiang, F. Bijlaard, and H. Kolstein. 2016. "Mechanical properties and design recommendations of very high strength steel S960 in fire." *Engineering Structures* 112, 60–70.

Ranawaka, T., and M. Mahendran. 2009. "Experimental study of the mechanical properties of light gauge cold-formed steels at elevated temperatures." *Fire Saf. J.* 44(2), 219–229.

Rothman, M., ed. 1988. *High temperature property data: Ferrous alloys*. Materials Park, OH: ASTM International.

Ruge, J., and O. Winkelmann. 1978. "Deformation behaviour of reinforcing and structural steel at high temperatures." Lund, Sweden: Lund Institute of Technology, Division of Building Fire Safety and Technology.

Sakumoto, Y., T. Yamaguchi, M. Ohashi, and H. Saito. 1992. "High-temperature properties of fire-resistant steel for buildings." *J. Struct. Eng.* 118(2), 392–407.

Sakumoto, Y., and H. Saito. 1995. "Fire-safe design of modern steel buildings in Japan." *J. Constr. Steel Res.* 33(1–2), 101–123.

Schneider, U. 1988. "Concrete at high temperatures—A general overview." *Fire Saf. J.* 13(1), 55–68.

Schneider, R, and J. Lange. 2010. "Constitutive equations and empirical creep law of structural steel S460 at high temperatures." *Proc., Structures in Fire*. Lancaster, PA: DEStech Publications.

Seif, M., J. Main, J. Weigand, F. Sadek, L. Choe, C. Zhang, et al. 2016a. "Temperature-dependent material modeling for structural steels: Formulation and application." *NIST Tech. Note 1907*. Gaithersburg, MD: NIST.

Seif, M., J. Main, J. Weigand, T. McAllister, and W. Luecke. 2016b. "Finite element modeling of structural steel component failure at elevated temperatures." *Structures* 6, 134–145.

Sha, W., F. Kelly, and Z. Guo. 1999. "Microstructure and properties of Nippon fire-resistant steels." *J. Mater. Eng. Perform.* 8(5), 606–612.

Shakya, A., and V. Kodur. 2016. "Effect of temperature on the mechanical properties of low relaxation seven-wire prestressing strand." *Constr. Build. Mater.* 124(10), 74–84.

Sidey, M., and D. Teague. 1988. "Elevated temperature data for structural grades of galvanized steels." *Tech. Note No. WL/PM/TN/251/88/D*. Port Talbot, UK: Welsh Laboratory, British Steel.

Skinner, D. 1972. "Measurement of the high temperature properties of steel." *Rep. No. MRL 6/10*. Melbourne, Australia: BHP Melbourne Research Laboratories, Broken Hill Proprietary.

Smith, C., B. Kirby, D. Lapwood, K. Cole, A. Cunningham, and R. Preston. 1981. "The reinstatement of fire damaged steel framed structures." *Fire Saf. J.* 4(1), 21–62.

Tao, Z., X. Wang, and B. Uy. 2013. "Stress–strain curves of structural and reinforcing steels after exposure to elevated temperatures." *J. Mater. Civ. Eng.* 25(9), 1306–1316.

Tide, R. 1998. "Integrity of structural steel after exposure to fire." *AISC Eng. J.* 35(1), 26–38.

Wan, R., F. Sun, L. Zhang, and A. Shan. 2011. "Study on microstructure and properties of low-Mo fire-resistant steel." *Adv. Mat. Res.* 168–170, 1792–1795.

Wei, Y., L. Zhang, F. Au, J. Li, and N. Tsang. 2016. "Thermal creep and relaxation of prestressing steel." *Constr. Build. Mater.* 128(12), 118–127.

Yu, L. 2006. "Behavior of bolted connections during and after a fire." Ph.D. dissertation, Austin, TX: Univ. of Texas, Dept. of Civil Engineering.

Yu, L., and K. Frank. 2009. "Shear behavior of A325 and A490 high-strength bolts in fire and post-fire." *AISC Eng. J.* 46(2). Accessed May 31, 2018. https://ascelibrary.org/doi/abs/10.1061/(ASCE)MT.1943-5533.0000445

Zhao, B., J. Kruppa, C. Renaud, M. O'Connor, and E. Mecozzi. 2005. "Calculation rules of lightweight steel sections in fire situations." *Final Rep., Contract No. 7210-PR/254.* Luxembourg: European Communities.

CHAPTER 9
STRUCTURAL ANALYSIS FOR FIRE EFFECTS

This chapter presents structural analysis methods for SFED, including finite element analysis. For SFED, simple calculations may be used for components and members, but they are discouraged for subsystems or systems because thermally induced loads due to elevated temperatures and restraint of thermal expansion are difficult to model with simple methods. Finite element analysis (FEA) methods with implicit or explicit solvers are available, both of which can be used to simulate temperature-dependent material properties and nonlinear behavior.

9.1 LOAD COMBINATIONS

Structural systems are designed using the applicable load combinations in ASCE 7-16, Section 2.3 or 2.4 (ASCE 2016). ASCE 7-16 Section 2.5 provides load combinations for checking structural performance for extraordinary loads (including fire exposure) that may lead to instability or collapse. Evaluation of structural members and elements for fire effects in the strength domain can be expressed quantitatively as

$$R_n(T) > \Sigma Q_{ni}(g) + Q_{ni}(T) \tag{9-1}$$

where

$R_n(T)$ = temperature-dependent member resistance,

$Q_{ni}(g)$ = gravity loads, and

$Q_{ni}(T)$ = temperature-dependent (thermally induced) loads.

Member resistance and loads may vary during a structural design fire. To obtain a desired level of structural reliability, resistance factors, φ, and load factors, γ_i, where i is applied to the design values for nominal member resistance, R_n, and member loads, Q_{ni}, are given as

$$\varphi R_n > \Sigma \gamma_i Q_{ni} \tag{9-2}$$

Resistance factors, φ, and load factors, γ_i, are developed to achieve the target reliability level as described in ASCE 7-16 Section C2.3.6 (ASCE 2016). Resistance factors are provided by material specifications.

For extraordinary events, load combinations (LC) 2.5-1 and 2.5-2 in ASCE 7-16 are used to evaluate the structural system performance (ASCE 2016).

$$(0.9 \text{ or } 1.2)D + A_k + 0.5L + 0.2S \qquad \text{(LC 2.5-1)} \tag{9-3}$$

$$(0.9 \text{ or } 1.2)D + 0.5L + 0.2(L_r \text{ or } S \text{ or } R) \qquad \text{(LC 2.5-2)} \tag{9-4}$$

For an uncontrolled fire, A_k is the thermally induced load in structural members and elements owing to elevated temperatures, D is the dead load, L is the live load, S is the snow load, L_r is the roof live load, and R is the rain load. Note that the load factor for A_k is 1.0 (Ellingwood and Corotis 1991, Ellingwood and Dusenberry 2005). These load combinations were developed for extraordinary events, such as an uncontrolled fire that may lead to ultimate limit states, such as gross inelastic deformation or partial collapse. In addition, the lateral stability of the structural system must be confirmed.

The live load factor of 0.5 and snow load factor of 0.2 in ASCE 7-16 and shown in Equation 9-3 is intended for typical occupancies where the live loads and snow loads at the time of the fire exposure is likely to be substantially less than the design loads (ASCE 2016). However, for occupancies where it is reasonable to conclude that the live load expected to be in place at the time of fire exposure is close to the design live load at ambient conditions, such as a warehouse, a live load factor greater than 0.5 may be appropriate. Some examples of occupancies where larger fire-exposed live load factors may be appropriate include dedicated storage areas, library stacks, and heavy equipment rooms. However, load factors should be developed in conjunction with target reliabilities and resistance factors using a standardized procedure, as described in ASCE 7 Section C2.3.6 (ASCE 2016). Such considerations should be memorialized within a design brief document prior to conducting analyses.

In ASCE 7-16 Section 4.7, live load reductions are not permitted when the live load is greater than 100 psf, except that members supporting two or more floors may be reduced by 20% (ASCE 2016). Minimum live loads for storage occupancies, library stacks, and manufacturing in ASCE 7-16, Table 4-1, are greater than 100 psf. Other codes, such as the National Building Code of Canada and the Eurocodes, prescribe live load factors for fire exposure that can be as much as 0.9 or 1.0 for certain storage occupancies. Furthermore, the Eurocodes consider wind load in conjunction with fire effects, which would have implications for the design of lateral members.

9.2 STRUCTURAL ANALYSIS TOOLS

9.2.1 Mapping Thermal Results to the Structural Model

Temperature histories of structural members and connections are needed to analyze the response of a structural system to fire exposure. The temperature histories depend on the fire characteristics (e.g., duration, heat release rate, fuel or ventilation limits, and so forth) and exposure scenario (e.g., proximity to fire, enclosure characteristics, and so forth). A time history of the thermal response (e.g., changes in material temperature) may include gradients across a section or along a member length.

Temperature history results from a thermal model typically need to be mapped to the structural model because the mesh size and element types will typically differ. Careful consideration should be given to the trade-off between optimal model features for each analysis versus mapping of temperatures between two sets of nodal data. The effects of using varying levels of discretization between the heat transfer and structural models should be considered. For example, the designer should consider the ramifications of using a heat transfer model that does not include thermal gradients (such as the lumped mass method, which computes uniform temperatures) to obtain temperature histories for a structural model composed of shell elements capable of simulating thermal gradients. Conversely, results of a 2D finite element heat transfer analysis would need to be simplified for use in a structural model with elements that do not include thermal gradients. However, SAFIR is an example of finite element software that seamlessly transitions from a thermal to a structural analysis because it was developed to conduct both thermal and structural analyses (Franssen and Gernay 2017). Alternatively, the designer may automate the mapping process with techniques such as tailored scripting.

Finite element models for heat transfer analyses typically use solid elements. Sections that are characteristically thin, such as steel beam flanges or webs, require finite elements that are quite small relative to the structural system. Structural finite element models typically use beam and shell elements, although solid elements may be used for a high-fidelity analysis of a subsystem or component of a structure. Element type and mesh discretization affect a number of modeling issues, including model size, quality of analysis results, convergence rates in implicit analyses, and the maximum time step in explicit analyses. For instance, a single beam element would represent a constant temperature along its length based on averaging of nodal input values. In most cases, results of finite element analyses of heat transfer will require mapping from a much denser mesh of solid elements to significantly fewer beam and shell elements comprising the structural model. Granted, the structural model should have an appropriate mesh density and number of section integration points to capture the pertinent structural behavior, such as large deflections and material plasticity.

Currently, there is no standardized method to conduct temperature mapping, and such efforts will depend on the skill and experience of the designer. In all cases, the designer should examine the process of temperature mapping for reasonableness. Accordingly, the thermal results should be examined to determine the general rate of temperature changes in structural components. The rate of change provides a basis for *snapshots* in time of temperatures across the structural system. Large temperature gradients typically require more careful mapping compared to more uniform temperatures.

9.2.2 Simplified Analytical Tools

Simplified methods of analysis may be sufficient when the scope of the SFED is limited to members and components rather than to systems. Also, the use of simplified methods should not preclude a designer from analyzing the performance of structural connections in any case. Analytical methods applicable to SFED are available in a few standards and guidelines. For example, AISC 360 Appendix 4 provides equations for determining the strength of simply supported members in flexure, compression, and tension at elevated temperatures (AISC 2016). However, other standards and guidelines are based on equivalence methods that are only applicable to SFRD (see Section 2.2.2) and do not address structural performance under realistic fire exposure.

Notable standards and guidelines for analytical methods applicable to SFED include the following:

- AISC 360 Appendix 4 has a section titled Simple Methods of Analysis, which addresses the performance of individual members during fire exposure. These methods assume that restraint conditions are unchanged from ambient temperature conditions. Equations are provided for calculation of member strength for structural design fires that address tension, compression, and flexural steel members, as well as composite floors (AISC 2016).
- *AISC Steel Design Guide 19* provides simplified equations for individual steel members subjected to fire exposure. For example, equations for the calculation of positive and negative flexural strengths in composite floors during fire exposure (AISC 2003).

Notable standards and guidelines based on equivalence methods that are only applicable to SFRD include the following:

- ASCE/SFPE 29 presents simple analysis methods for a variety of structural components and assemblies to derive a fire resistance rating (ASCE/SFPE 2005).
- ACI 216.1 provides simplified procedures to calculate the fire resistance rating of concrete and masonry members. This standard states that its design and analysis procedures shall be based on the standard fire exposure and applicable acceptance criteria of standard fire testing. Notably, Section 2.4 of the standard provides concrete cover requirements required to achieve certain levels of fire resistance (ACI 2007).

The Eurocodes provide guidance for SFED (see Phan et al. 2010 for further information). The following Eurocode guidance documents address SFED of steel and concrete structural members and composite systems:

- Eurocode 3 provides simplified calculation methods for evaluating steel members under fire exposure. However, the designer should not use equations or calculation methods in this code that pertain to fire resistance ratings, which are only applicable to SFRD. For example, this code contains modified steel member fire resistance rating values for a variety of load types, including tension, compression, and combined bending and axial compression (CEN 2005a).
- Eurocode 2 provides simplified calculation methods for evaluating concrete members subjected to fire exposure. However, the designer should not use equations or calculation methods in this code that pertain to fire resistance ratings, which are only applicable to SFRD. For example, temperature profiles are provided for member cross sections exposed to the standard fire, which is only applicable to SFRD (CEN 2004).
- Eurocode 4 provides simplified calculation methods for evaluation of composite systems subjected to fire exposure (CEN 2005b). Similar to Eurocode 2, the designer should not use equations or calculation methods in this code that pertain to fire resistance ratings, which are only applicable to SFRD.

9.2.3 Finite Element Analysis

FEA has become an increasingly effective tool for SFED, especially for analysis of steel structures. FEA can simulate the performance of a structural system under fire exposure. Full-scale structural fire testing of large assemblies is uncommon because of the costs and time required to construct and conduct such testing. Hence, FEA is used in many cases to determine structural response to fire exposure by accounting for the framing layout, loading, and boundary conditions. This section discusses recommended FEA modeling practices for SFED and reviews some of the available finite element analysis software.

The complexity of structural behavior under fire exposure can vary from degraded material properties of isolated members to thermally induced interactions of heated structural assemblies and adjacent structural framing. The loss of member load-carrying capacity under fire exposure may activate load redistribution mechanisms, where the load that is shed by the failed member is redistributed to the adjacent members. FEA of system-level behavior of structures subjected to fire effects should capture the local behavior of the individual structural members, as well as the structural interaction and load redistributions.

An FEA model can be used to identify alternative load paths that may be activated if failure of certain structural members occurs, as well as possible progressive failures. Robust finite element modeling approaches must, at a minimum, include the nonlinear stress–strain relationships of materials and temperature data for elements, including uniform and gradient temperature profiles. A suitable finite element model should appropriately address the relevant facets of structural response under fire exposure. Examples of typical behavior of structural components exposed to fire effects include but are not limited to the following:

- **Steel framing:** Increased demands, plastic response with strain hardening, local buckling, lateral torsional buckling, distortional buckling, geometric nonlinear large deflection behavior (e.g., catenary action), and modified material properties.
- **Concrete or masonry members:** Increased demands, plastic response with strain hardening, cracking, localized crushing, loss of section due to spalling, and modified material properties.
- **Wood members:** Increased flexural bending and shear, tensile rupture, longitudinal splitting, loss of section due to charring, and modified material properties.
- **Connections:** Varying force and moment due to restraint of thermal expansion during heating and cooling phases of fire exposure.
- **Slabs:** Thermally induced compressive actions and tensile membrane actions due to modified material properties.

FEA software packages offer a wide selection of element formulations for which a particular structural member or system may be modeled. Structural element types may include beam, shell (e.g., Figure 9-1), and solid elements, and connector-type elements (e.g., springs). Beam-type elements are often used to model global structural behavior. In SFED, it is sometimes necessary for a more detailed understanding of a member's local behavior. In these situations, a beam-type element may not suffice, and the use of higher-order elements (e.g., shell or solid elements) may be necessary (e.g., as shown in Figure 9-1). In all cases, the structural element type chosen

Figure 9-1. Shell element representation of a girder-to-column connection. Source: Simpson Gumpertz & Heger Inc.

should be able to accommodate the anticipated temperature profiles. Typically, the selection of element types is guided by the anticipated complexity of the structural response in the model.

There are both general purpose and specialty commercial FEA software available for application to SFED. Generally, available commercial FEA software (e.g., Abaqus) has been adapted for SFED in many cases. Also, there is FEA software that has been specifically developed for SFED of steel and composite structures by universities and research organizations. A review of the capabilities of SFED-dedicated thermal and structural analysis programs is provided by Terro (1991) and Wang (2002). Table 9-1 lists legacy and contemporary FEA software with SFED capabilities.

FEA software packages can use several analysis options depending on the problem:

- *Static vs. Dynamic*: Static analyses neglect damping and inertial effects, which may be appropriate for some SFED analyses. Dynamic analyses include nodal forces associated with damping and inertial effects. A dynamic analysis should be considered if the acceleration or velocity

Table 9-1. SFED Finite Element Analysis Software.

Program	Description	Author/ Developer	Year
Legacy Software			
FIRES-T3	Computer program for the fire response of structures—Thermal cases	Iding et al.	1977
FIRES-RCII	Computer program for the fire response of structures—Reinforced concrete frames.	Iding et al.	1977
TASEF	Computer code for calculating the temperature in structures exposed to fire	Wickstrom	1979
CEFICOSS	Computer program for the fire design of composite and steel structures	Schleich et al.	1986
STABA-F	Computer program for the determination of load-bearing and deformation behavior of uniaxial structural elements under fire	Rudolph et al.	1986
FASBUS-II	Fire analysis of steel building systems	Jeanes	1985
USFOS	Computer program for nonlinear static and dynamic analysis of space frame structures.	SINTEF	1988
Contemporary Software			
FEAST	Finite element analysis of shells at high temperature	Gillie et al.	2000
ADAPTIC	Adaptive static and dynamic structural analysis program	Izzuddin	2000
SAFIR	Special purpose computer program for the analysis of structures subjected to fire	University of Liege, Belgium	2005
Opensees	Special purpose computer program for the analysis of structures subjected to earthquake, and specifically extended to analysis of fire effects (Jiang and Usmani 2013)	University of California Berkeley	2013
Vulcan	Finite element analysis program capable of modelling the global 3-dimensional behavior of composite steel-framed buildings under fire conditions	University of Sheffield and Vulcan Solutions	2016
ABAQUS	Generally available commercial software package (implicit and explicit)	Dassault Systèmes	2016

Program	Description	Author/ Developer	Year
Contemporary Software			
DIANA	Generally available commercial software package	DIANA FEA BV	2016
Ansys, LS-Dyna	Generally available commercial software package (implicit and explicit)	ANSYS, Inc.	2016

Disclaimer: Certain trade names or company products are mentioned in the text. In no case does such identification imply recommendation or endorsement, nor does it imply that the products are the best available for the purpose.

of nodes becomes non-negligible. Static analyses use an implicit solver, whereas dynamic analyses can use either an explicit or implicit solver.

- *Implicit vs. Explicit:* Implicit FEA solvers perform iterative calculations for each time step until convergence is achieved (i.e., when the changes between iterations become sufficiently small). Most implicit numerical solvers invert the stiffness matrix for each calculation, and it can be an expensive operation, especially for large models. Explicit FEA solvers use small time-step increments and calculate a solution based on the state of the system from the previous time step. Explicit time steps must be less than the Courant time step (the time it takes a sound wave to travel across an element). In an explicit analysis, no iteration is required, as the nodal accelerations are solved directly.

Implicit FEA solvers are generally more efficient in computational time than explicit solvers if the analysis does not include significant nonlinearities in the material properties or structural behavior. Implicit analyses also require specialized numerical techniques to simulate sequential failures in a structural system. Explicit analyses can address structural and material nonlinearities and sequential failures with relative ease compared to implicit analyses as nodal accelerations are solved directly (not iteratively) as the inverse of the diagonal mass matrix times the net nodal force vector. The quality of explicit solutions is based on the model mesh discretization and minimum time step. Explicit solvers can readily handle instabilities all the way to collapse, but the designer must exercise caution with the rate of load application to minimize artificial inertial effects. However, it is often unfeasible to validate explicit results.

When using FEA, the material stress–strain behaviors are input for the temperature range of interest, as derived from thermal response analyses. The designer should be aware that the representation of these properties can be either uniaxial or multiaxial. Strains can generally be linearly interpolated between the supplied stress–strain curves. Each stress–strain curve requires a linear modulus of elasticity up to the proportional limit and/or yield point and a series of plastic stress–strain points up to the ultimate strain (McAllister et al. 2008). The elastic modulus and stress–strain points at yield must be consistent for each selected temperature.

Significant reductions in computational time can be achieved with judicious selection of FEA element types and mesh discretization. It is the designer's responsibility to conduct appropriate convergence studies or provide references to justify the level of discretization used for each analysis, which may vary depending on the application. The key with any element selection is to reduce computational effort while having a negligible impact on computational accuracy. On the basis of minimum computational time, beam elements may be a better choice for a system-level analysis. For detailed analysis of certain structural members, the typical trade-off is between a shell and a solid element, and a common metric for this decision is the slenderness ratio of the structural member (i.e., the ratio of a structural member's length with respect to its thickness). The smaller the slenderness ratio, the more computationally accurate it becomes to select a solid versus a shell element. For example, a built-up steel plate girder may have a web slenderness ratio (i.e., depth of the web divided by the thickness of the web) of 150, which indicates that the depth of the web is 150 times the thickness of the web. At this large slenderness ratio value, the web may typically be modeled with shell elements for a reduction in computational effort without sacrificing computational accuracy.

There is no well-defined decision point for the slenderness ratio value at which a finite element model should use a shell versus a solid element. However, the designer should consult benchmark models, such as those developed by the European Cooperation in Science and Technology (COST 2014), which are specifically provided for a particular FEA software package. These benchmark models allow the designer to check FEA accuracy against a well-documented and verified set of accepted performance values. These benchmark models also allow the user to check the influence that altering a particular parameter may have on the final result. For example, depending on the shell element that is selected, the number of integration points through the element thickness may be adjusted depending on the level of accuracy that is necessary through the element thickness. FEA models ultimately should be validated against published, accepted values (e.g., published experimental values or benchmark models), and it is the responsibility of the designer to check that the FEA model is a reasonably accurate representation of physical reality.

9.3 STRUCTURAL ANALYSIS METHODS

9.3.1 Primary Structural Framing

The stability of structures under fire exposure depends on the system-level performance of primary structural framing (McAllister et al. 2008, Agarwal et al. 2009, Agarwal et al. 2014, Varma et al. 2008). As strength and stiffness of structural members and elements degrade with increasing temperature, the structural system may become vulnerable to collapse. In particular, excessive deflection of floor beams under fire exposure and failure of connections may render columns unbraced for more than one story height and increase their vulnerability to buckling failure.

Under fire exposure, columns may experience rotational and axial restraints from cooler portions of the building above and below the heated area (Agarwal et al. 2009, 2011, 2014; Choe et al. 2011; Choe et al. 2017). These restraints influence the capacity of the columns (Kodur and Agarwal 2016) and change the boundary conditions from those at ambient temperature. AISC 360 provides guidance on how to adjust the effective length of a column to determine the column's temperature-dependent load-carrying capacity (AISC 2016).

An important decision to make when modeling column behavior is whether temperature gradients through the cross section of the column and along the height of the column are significant to the member performance. Although many designers may assume a uniform temperature gradient along the height of the column, the same assumption through the cross section can lead to an overestimation of column capacity. Equation A-4-2 in AISC 360 Appendix 4, which is based upon research developed by Takagi and Deierlein (2007), can be used to calculate the temperature-dependent column capacity for the inelastic flexural buckling failure mode. Similarly, the equation developed by Agarwal and Varma (2011) is based on calibrated temperature-dependent yield stress and elastic modulus values for steel. Lastly, the column capacity equation proposed by Talamona et al. (1997) considers columns that are not uniformly heated such as perimeter columns that may be only heated from one side. Previous researchers have demonstrated that when only one side of the column is exposed to fire, assuming a uniform temperature distribution through the cross section can grossly overestimate the column capacity (Olawale 1988, Garlock and Quiel 2008, Antonio et al. 2010, Seif et al. 2016).

When a column is not heated uniformly around the perimeter during fire exposure, the hotter side expands more rapidly than the cooler side. If the column were modeled by itself with idealized pinned ends, this gradient would cause thermal bowing in the column toward the hotter side. However, realistic rotational stiffness due to continuity at the column ends will restrain the bowing and induce a moment that increases the compression in the hotter side. At the same time, asymmetric heating can shift the neutral axis of the cross section toward the cooler side of the column cross section as the mechanical properties in the hotter side decrease more rapidly. This shift, coupled with the axial force on the column, causes a secondary moment that leads to an increase in compression on the cooler side of the column. The interaction of these moments with the applied axial force will contribute to the column's fire-induced performance (Garlock and Quiel 2008). Such behavior has been confirmed through experimental tests of columns subjected to thermal gradients (Agarwal et al. 2014). Accordingly, the designer should consider these possible column behaviors due to thermal gradients when conducting an SFED.

9.3.2 Composite Slabs and Floor Systems

Under fire exposure, floor beams may experience large deflections. The beam load may be redistributed to the slab and surrounding floor system. Therefore, analysis of the composite slab is crucial for determining realistic loads in composite floor systems.

During an enclosure-type fire, the surrounding cooler portions of a structural system may provide restraint against thermal expansion of the floor beams. Research has shown that analysis of fire effects on a floor system can provide insight into more realistic failure modes of the composite beams and connections as opposed to modeling a single bay (McAllister et al. 2012, Fischer and Varma 2017, Garlock and Selamet 2010). When evaluating the performance of a structural system under fire exposure, consideration of only one bay of the structural system may result in an overly conservative estimate of the performance and may demonstrate failure modes that would not occur when considering the overall floor system. Agarwal and Varma (2014) demonstrated that in the event of column failure within a building, the floor system may be able to redistribute the load to the surrounding columns. Conversely, irregularity in the structural system layout can induce additional forces and deformations in primary members and composite floor assemblies under fire exposure and unexpected failure mechanisms in connections (Flint et al. 2013).

Composite floor slabs may experience tensile or compressive membrane action during fire exposure. Tensile membrane action is a mechanism where floor slabs may undergo large vertical displacements or where induced radial tension in the center of the slab is resisted by a ring of compression around the periphery of the floor slab. A vertical deflection at least equal to the slab thickness marks the beginning of this mechanism. Necessary conditions include a slab designed for two-way bending, vertical support at all four edges, and an aspect ratio of 2:1 or less. Compressive membrane action occurs when thermal expansion of the slab is restrained, and it can also increase the slab performance during fire exposure. An overview of current analysis methods for tensile membrane action in composite slabs for SFED is given in Buchanan and Abu (2017). However, tensile membrane action is only effective if the supporting edge beams do not undergo large deformations.

To simulate composite beams, either FEA or fiber-based models can be used for SFED. Fiber-based models consider the structural cross section to be divided into strips or *fibers* that are extruded to create beam or shell elements. Each fiber typically has temperature, stress, and strain that are subjected to structural equilibrium conditions during each step of a computational analysis. Both fiber-based and FEA methodologies can simulate the section-level moment-curvature behavior of composite beams. All materials within the model should account for temperature-dependent stress–strain relationships, and load-slip behavior at the steel–concrete interface of the composite beams. Section 9.3.3 discusses the representation of shear connectors. AISC 360 Appendix 4 (AISC 2016) permits the inclusion of membrane action in composite floor slabs if the design accounts for the increased tensile forces on the connections, as well as gravity load redistribution to the adjacent framing. Section 4.2.4d.d in AISC 360 Appendix 4 also provides guidance for simplified design of composite beams for flexure.

In certain cases, steel beams (such as castellated beams) are installed with cut-out openings for mechanical systems along the length of their webs. Designers should consider the effect that such openings have on the structural capacity of the beams under fire exposure and consider failure modes related to the holes and web slenderness, such as overall vertical shear, overall bending moment, vertical shear at openings, bending moment at openings, Vierendeel bending moment at openings, web-post buckling, web-post bending, and web-post horizontal shear (SCI 2011). In these cases, beam failure under fire exposure would be typically associated with the web-posts, which essentially act as a series of columns or struts within the beam itself (between openings). Also, failure modes generally occur at lower temperatures as compared to a plain beam of the same size because failure typically relates to buckling rather than bending. For beams with web openings, the designer should conduct SFED with consideration of the specific failure modes described.

9.3.3 Structural Connections

Structural design for ambient conditions typically uses linear analyses with P-delta effects, and structural framing connections are often idealized as either fully rigid (full bending moment transfer that maintains original angles between members) or simple (no bending moment transfer). Also, framing members and connections are designed such that failures are expected to occur in framing members before connections and, therefore, connections are modeled to transmit forces and moments without failure modes.

Structural connection design for ambient conditions accounts for gravity loads and lateral loads (e.g., due to wind and seismic effects), but they do not account for thermally induced axial loads and flexure during fire exposure. When evaluating a structural system for fire effects, idealized connection models may not be sufficient, as thermally induced failures may occur in connections before they occur in framing members. At a minimum, the designer should check thermally induced connection forces and moments against the thermally reduced capacity of the connection. Both the Cardington tests (British Steel 1999) and the World Trade Center studies (McAllister et al. 2012) indicate that high-fidelity models of connection behavior may be needed to fully evaluate the structural system response.

Steel columns and connections in the Cardington tests were thermally protected during some tests with the floor sections left unprotected. Failure mechanisms observed during the heating phase were primarily local buckling of lower beam flanges where adjacent, cooler framing restrained the thermal expansion of the beams. Failure mechanisms observed during the cooling phase were connection failures due to bolt shear rupture or plate fracture.

High-fidelity models of floor framing connections were developed as early as the 1990s. For example, an FEA element was developed to represent semi-rigid connection behavior with a moment-rotation relationship (Wang and Moore 1995). Also, the performance of single-plate shear (fin) connections under fire exposure was studied by Sarraj et al. (2007) and Garlock and Selamet (2010). More recently, detailed models of floor framing connections that include failure modes were developed for the WTC 7 studies (McAllister et al. 2012). The floor framing connection models included all possible failure modes, such as bolt shear rupture, weld rupture, and bearing or tear-out failure of the shear plate. The connection models used temperature-dependent material properties for all components, including the bolts, which lose strength more quickly than typical structural steels.

Shear stud failures may occur if there is differential thermal restraint between the slab and floor beams. In buildings, floor slabs are often continuous, whereas the floor beams frame into girders or columns with connections that provide less thermal restraint. For the WTC 7 studies (McAllister et al. 2012), the shear stud connections were modeled with an empirical relationship that incorporates multiple failure modes, including shear cone failure in the concrete slab and shear stud weld failure at the beam flange. The strength and stiffness of the shear studs can be simulated using previous work performed by Oehlers and Sved (1995) and Mujagic and Easterling (2009). The temperature-dependent load-slip relationship of the shear stud at the steel–concrete interface can be simulated using research from

Zhao and Kruppa (1997). Some researchers have demonstrated that the load-slip behavior can be assumed to be evenly distributed along the length of the beam (Selden and Varma 2016, Fischer and Varma 2017).

Simplified connection models have been developed using nonlinear springs for single-shear plate connection failure modes under tensile forces (Sarraj et al. 2007, Sadek et al. 2008, Weigand 2017). The nonlinear spring approach offers a reasonable middle ground between idealized connections and detailed FEA models, and they may be more conducive to SFED generally. Designers should select a level of model fidelity that would reasonably and conservatively represent the anticipated structural behaviors.

9.3.4 Structural Walls

During an uncontrolled fire event, structural walls are often exposed on only one side, and thermal gradients typically develop through wall thickness. Depending on the construction material and the wall thickness, these gradients can significantly affect the structural performance of the wall. Thermal gradients can induce out-of-plane bending moments due to the restraint of thermal expansion through the wall thickness. The wall's neutral axis may shift toward the cooler face that is not exposed to fire. If this occurs, vertical loads applied through the geometric centroid of the wall may induce an eccentric out-of-plane moment. The combination of these moments can lead to permanent damage, as well as catastrophic failure (Mueller and Kurama 2015, 2017).

SFED of structural walls should account for the potential development of a thermal gradient through the wall thickness, as well as the associated restraint of nonuniform thermal expansion and nonuniform material weakening. For characteristically thin walls composed of steel or other metals, the designer can typically neglect the development of a thermal gradient through the wall thickness.

REFERENCES

ACI (American Concrete Institute). 2014. *Code requirements for determining fire resistance of concrete and masonry construction assemblies*. ACI 216.1-14/TMS 0216.1-14. Farmington Hills, MI: ACI.

ACI/TMS. 2007. *Code Requirements for Determining Fire Resistance of Concrete and Masonry Construction Assemblies*. ACI 216.1-07/TMS-216-07. Farmington Hills, MI: ACI.

AISC (American Institute of Steel Construction). 2003. *Design guide 19: Fire resistance of structural steel framing*. Chicago: AISC.

AISC. 2016. *Specification for structural steel buildings*. ANSI/AISC 360. Chicago: AISC.

Agarwal, A., A. H. Varma, and G. Cedeno. 2009. "Steel columns under fire loading: Stability behavior and design." *Proc. Ann. Stability Conf., Structural Stability Council*, Phoenix, AZ.

Agarwal, A., and A. H. Varma. 2011. "Design of steel columns at elevated temperatures due to fire: Effects of rotational," *Eng. J.* 48, 297–314.

Agarwal A., and A. H. Varma. 2014. "Fire-induced progressive collapse of steel building structures: The role of interior gravity columns." *Eng. Struct.* 58, 129–140.

Agarwal, A., L. Choe, and A. Varma. 2014. "Fire design of steel columns: Effects of thermal gradients." *J. Constr. Steel Res.* 93, 107–118.

Antonio M., P. Joao, and P. Valdir. 2010. "Numerical study on the behavior of steel columns embedded on brick walls subjected to fire." *Proc., Int. Symp. Steel Structures: Culture and Sustainability*, Turkey.

ASCE. 2016. *Minimum design loads for buildings and other structures*. ASCE/SEI 7. Reston, VA: ASCE.

ASCE/SFPE (Society of Fire Protection Engineers). 2005. *Standard calculation methods for structural fire protection*, ASCE/SFPE 29. Reston, VA: ASCE.

British Steel. 1999. "The behavior of multi-storey steel framed buildings in fire." South Yorkshire, UK: Swinden Technology Centre, A European Joint Research Program.

Buchanan, A., and A. Abu. 2017. *Structural design for fire safety*, 2nd Ed. New York: Wiley.

CEN. 2004. *Eurocode 2—Design of concrete structures. Part 1–2: General rules—Structural fire design*. EN 1992-1-2. Brussels: CEN.

CEN. 2005a. *Eurocode 3—Design of steel structures. Part 1–2: General rules—Structural fire design*. EN 1993-1-2. Brussels: CEN.

CEN. 2005b. *Eurocode 4—Design of composite steel and concrete structures. Part 1–2: General rules—Structural fire design*. EN 1994-1-2. Brussels: CEN.

Choe, L., A. Varma, A. Agarwal, and A. Surovek. 2011. "Fundamental behavior of steel beam-columns and columns under fire loading: An experimental evaluation." *J. Struct. Eng.* 137(9), 954–966.

Choe, L., C. Zhang, W. E. Luecke, J. L. Gross, and A. H. Varma. (2017). "Influence of material models on predicting the fire behavior of steel columns." *Fire Technol.* 53(1), 375–400.

COST (European Cooperation in Scientific and Technology). 2014. *Benchmark studies: Verification of numerical models in fire engineering*. Prague: CTU Publishing House.

Ellingwood, B. R., and R. B. Corotis. 1991. "Load combinations for buildings exposed to fires." *Eng. J.* 28, 37–44.

Ellingwood, B., and D. Dusenberry. 2005. "Building design for abnormal loads and progressive collapse." *Comput. Aided Civ. Inf. Eng.* 20(3), 194–205.

Fischer, E., and A. Varma. 2017. "Fire Resilience of Composite Beams with Simple Connections: Parametric Studies and Design." *J. Constr. Steel Res.* 128, 119–135.

Flint, G., S. Lamont, B. Lane, H. Sarrazin, L. Lim, D. Rini, and C. Roben. 2013. "Recent lessons learned in structural engineering for composite steel structures." *Fire Technol.* 49(3), 767–792.

Franssen, J., and T. Gernay. 2017. "Modeling structures in fire with SAFIR: Theoretical background and capabilities." *J. Struct. Fire Eng.* 8(3), 300–323. DOI: 10.1108/JSFE-07-2016-0010.

Garlock M., and S. Quiel. 2008. "Plastic axial load and moment interaction curves for fire exposed steel sections with thermal gradients." *J. Struct. Eng.* 134(6), 874–880.

Garlock, M., and S. Selamet. 2010. "Modeling and behavior of steel plate connections subject to various fire scenarios." *J. Struct. Eng.*, 136(7), 897–906. DOI: 10.1061/(ASCE)ST.1943-541X.0000179.

Jiang, J., and A. Usmani. 2013. "Modeling of steel frame structures in fire using OpenSees." *Computers & Structures* 118, 90–99.

Kodur, V., and A. Agarwal. 2016. "An approach for evaluating residual capacity of reinforced concrete beams exposed to fire." *J. Eng. Struct.* 107, 293–306.

McAllister, T., R. Gann, J. Averill, J. Gross, W. Grosshandler, J. Lawson, et al. 2008. "Federal building and fire safety investigation of the World Trade Center disaster: Structural fire response and probable collapse sequence of World Trade Center building 7." *NCSTAR 1–9*. Gaithersburg, MD: NIST.

McAllister, T., F. Sadek, J. Gross, S. Kirkpatrick, R. MacNeill, M. Zarghamee, et al. 2012. "Structural analysis of impact damage to World Trade Center buildings 1, 2 and 7." *Fire Technol.* 49(3), 615–642. DOI: 10.1007/s10694-012-0289-2.

Mueller, K., and Y. Kurama. 2015. "Out-of-plane behavior and stability of five planar reinforced concrete bearing wall specimens under fire." *J. Am. Concr. Inst.* 112(6) 701.

Mueller, K., and Y. Kurama. 2017. "Out-of-plane behavior of reinforced concrete bearing walls after one-sided fire." *J. Am. Concr. Inst.* 114(1), 149.

Mujagic, J., and W. Easterling. 2009. "Reliability assessment of composite beams." *J. Constr. Steel Res.* 65(12), 2111–2128.

Oehlers, D., and G. Sved. 1995. "Composite beams with limited-slip-capacity shear connectors." *J. Struct. Eng.* 10.1061/(ASCE)0733-9445(1995)121:6 (932).

Olawale, A. 1988. "Collapse behavior of steel columns in fire." Ph.D. thesis, Univ. of Sheffield, Sheffield, UK.

Phan, L., T. McAllister, J. Gross, and M. Hurley. 2010. *NIST TN 1681: Best practice guidelines for structural fire resistance design of concrete and steel buildings*. Gaithersburg, MD: NIST.

Sadek, F., S. El-Tawil, and H. Lew. 2008. "Robustness of composite floor systems with shear connections: Modeling, simulation, and evaluation." *J. Struct. Eng.* 134(11), 1717–1725.

Sarraj, M., I. Burgess, J. Davison, and R. Plank. 2007. "Finite element modeling of steel fin plate connections in fire." *Fire Saf. J.* 42(6-7), 408–415.

Seif, M., J. Main, J. Weigand, F. Sadek, L. Choe, C. Zhang, et al. 2016. *Temperature-dependent material modeling for structural steels: Formulation and application*. Gaithersburg, MD: NIST.

Selden, K., and A. Varma. 2016. "Flexural capacity of composite beams subjected to fire: Fiber-based models and benchmarking." *Fire Technol.* 52(4), 995–1014.

SCI (Steel Construction Institute). 2011. "Fire design of composite beams with rectangular and circular beam openings." *RT1356*, Ascot, UK: SCI.

Takagi, J., and G. Deierlein. 2007. "Strength design criteria for steel members at elevated temperatures. *J. Constr. Steel Res.* 63(8), 1036–1050.

Talamona, D., J. Franssen, J. Schleich, and J. Kruppa. 1997. "Stability of steel columns in case of fire: numerical modeling." *J. Struct. Eng.* 123(6), 713–720.

Terro, M.1991. "Numerical modelling of thermal and structural response of reinforced concrete structures in fire." Ph.D. thesis, Imperial College of Science and Technology, London.

Varma, A., S. Hong, A. Agarwal, and K. Prasad. 2008. "Behavior of steel building structures under realistic fire loading." *Proc., Structures Congress*. Reston, VA: SCI.

Wang, Y. 2002. *Steel and composite structures: Behaviour and design for fire safety*. Boca Raton, FL: CRC Press.

Wang, Y., and D. Moore. 1995. "Steel frames in fire: Analysis." *Eng. Struct.* 17(6), 462–472.

Weigand. 2017. "Component-based model for single-plate shear connections with pretension and pinched hysteresis." *J. Struct. Eng.* 143(2), 04016178. DOI: 10.1061/(ASCE)ST.1943-541X.0001662.

Zhao, B., and J. Kruppa. 1997. "Fire resistance of composite slabs with profiled steel sheet and composite steel concrete beams Part 2: Composite beams." *Final Report, ECSC—Contract No. 7210 SA 509*, Centre technique de la construction métallique (CTICM), France.

CHAPTER 10
STRUCTURAL ACCEPTANCE CRITERIA

10.1 OVERVIEW

Acceptance of a structural fire engineering design (SFED) is subject to approval by the building authority, which may require a third-party review. Acceptance criteria should reflect the intent of the minimum performance objectives as stipulated in ASCE/SEI 7-16 Appendix E (see Section 3.1), as well as those discretionary performance objectives that are enacted for a given project (if any) (ASCE/SEI 2016).

The acceptance criteria should be quantitative measures to fulfill the required performance objectives. Acceptance criteria should meet the minimum criteria specified in Sections 10.2 through 10.5, based on the structural system material type. It is recommended that salient acceptance criteria be agreed upon by the project stakeholders and documented in a design brief document, especially those that are not explicitly defined in Sections 10.2 through 10.5.

When undertaking an SFED, the designer should exercise appropriate skill, care, and acquired competency in determining and using specific acceptance criteria for a given structural system, especially those that are not explicitly defined in Sections 10.2 through 10.5. This may be facilitated by a third-party review at the discretion of the owner or building authority. It is the duty of the designer to confirm adequate load paths for all assemblies or elements and to ensure appropriate redundancy when considering each structural material for the hierarchy of failure modes (see Section 7.1). Each material will behave uniquely, and partial and full collapse should be considered.

10.2 STRUCTURAL STEEL

When assessing the adequacy of a steel structure to satisfy the required performance objectives under fire exposure, the designer should consider all relevant limit states and their consequences with respect to the required performance objectives, as described in the following sections. Limit states for local behavior and gross member behavior can impact the performance of steel structures under fire exposure.

10.2.1 Local Behavior

Local behavior may control the response of steel structures under fire exposure, even if they are not controlling limit states at ambient temperature. For example, it is sometimes reasonable and advantageous to design for large displacements of steel gravity systems under fire exposure. However, such an approach may introduce new limit states, which do not control the design at ambient temperature.

Limit states that may cause failure (i.e., loss of load-carrying capacity) in steel structures due to fire exposure may include local buckling of beam or column flanges, brittle fracture of highly stressed components, and ductile fracture of highly strained components. Currently, the available literature on the subject is not robust enough to define generalized acceptance criteria for each of these limit states for the full range of temperatures of interest in SFED. However, it has been demonstrated that sophisticated structural models coupled with calibrated mechanical properties (see Chapter 8) can provide reasonable predictions of local failures (Gernay et al. 2013, Gernay and Franssen 2015).

The use of local acceptance criteria for steel structures requires a commensurate level of sophistication in the structural analysis. Accordingly, an FEA model should provide sufficient refinement (e.g., mesh size) and an appropriate constitutive model (see Chapter 8) that is calibrated for the temperatures and stresses of interest. Structural elements in which local buckling is a critical limit state should be evaluated using closely related test data and an accepted constitutive model from Section 8.2. Local buckling in beams can be modeled as recommended by Seif et al. (2016).

Models for brittle and ductile fracture in steel at ambient temperature have been validated for SFED applications and are summarized in the review by Kanvinde (2016). These models have been demonstrated to be accurate at elevated temperatures (Johnson and Cook 1985). However, generalized acceptance criteria are not feasible given the wide variation in fracture toughness between nominally identical structural steels, even at ambient temperature (Myers et al. 2009). In cases where the structural analysis demonstrates fracture to be a critical limit state [e.g., local strains greater than 10% or greater than 2% in the presence of significant stress concentrations and tensile loads (Kanvinde 2016)], a dedicated fracture mechanics analysis that is supported by appropriate test data should be used by the designer.

10.2.2 Gross Member Behavior

10.2.2.1 Simplified Analytical Tools. When assessing strength and stability limit states using hand calculations, the designer should consider the change in behavior due to the indistinct yield point of steel at elevated temperatures. As stated in Section 9.1, the temperature-dependent member capacity must be greater than the demand due to gravity applied loads and thermally induced loads from fire exposure, as required to fulfill the required performance objectives. In cases where hand calculations suggest that the demand exceeds the capacity, it is recommended that the designer either redesign the structural member such that the capacity exceeds the demand or analyze the structural system using advanced methods of analysis, such as those described in Section 10.2.2.2. When using hand calculations, a demand-to-capacity ratio less than 1.0 indicates a conservative design. A demand-to-capacity ratio exceeding 1.0 requires further analysis, but it does not necessarily indicate that a structural system would become unstable during a structural design fire, and stability may be demonstrated with a more advanced analysis. Further research is required to understand the anticipated performance of *over-capacity* structural steel sections when considering demand-to-capacity ratios calculated with simplified analytical tools under fire conditions. As stated in Section 9.2.2, the use of simplified methods should not preclude a designer from analyzing the performance of structural connections in any case.

In addition, the designer should ensure that the structural analysis considers changes to lateral restraint during a fire (e.g., due to weakening of bracing elements). The designer should exercise engineering judgment to ensure that the assumptions and limitations of simple analytical tools remain valid for the temperatures expected during a structural design fire. Examples of conditions that may preclude the use of hand calculations include the occurrence of excessive secondary moments due to large axial forces from restrained thermal expansion coupled with large deformations, and the reliance on system-level response through methods such as tensile membrane action. These more complex phenomena cannot be accurately captured with hand calculations, and thus require the use of a more advanced method, as described in the following section.

10.2.2.2 Finite Element Analysis. Structural response to fire effects is dependent on geometric and material nonlinearities, and nonlinear FEA can be used to design for this behavior. The designer should perform applicable hand calculations for comparison to critical FEA results. It is recommended that such hand calculations follow the guidance provided in Section 10.2.2.1.

Under fire exposure, a structural system may reach a highly deformed state without failing (i.e., loss of load-carrying capacity). An acceptable criterion of L/20 has been proposed in the literature for overall structural stability (Wang et al. 2013). Along with other necessary capacity checks (e.g., at connections), this deflection limit is reasonable for SFED. In any case, the designer should consider this limit as a convenient benchmark and exercise engineering judgment when determining appropriate deflection limits for a given condition.

For structural members with applied insulation, the designer should properly contemplate the reliability of the insulation to remain intact during the fire (see Section 6.5). Accordingly, deflection limits (total deflection and rate of deflection) imposed by standard fire testing (within SFRD) may be referenced (Ryan and Robertson 1959, Bono 1963). However, these limits are based on the scope and acceptance criteria of a standard fire test and may not be representative of in-situ performance. Beyond these limits and for fire exposures significantly more intense than the standard fire, research and/or specific test data are required to understand the implications on applied insulation. Lacking specific information from the literature or the insulation manufacturer, SFPE S.02 (2015) recommends that applied insulation used for SFED be qualified by standard fire testing with a heating exposure that is more intense than the structural design fires under consideration. For example, a designer may choose to specify applied insulation that has been qualified to the UL 1709 (2011) hydrocarbon curve for conservatism because this test curve is more intense than most structural design fires that typically involve ordinary combustibles.

10.2.3 Connection Design

Under fire exposure, structural connections may be subjected to a complex variation of forces that almost certainly would not have been contemplated during design for ambient conditions. Accordingly, the designer should consider the effects of member deformations on the capacity of connections, including large deflections, catenary behavior, and thermal expansion. Axial thrust on structural connections from restrained thermal expansion combined with rotation due to large member displacement may result in connection failure (i.e., loss of load-carrying capacity). Figure 10-1 shows a steel beam cleat connection that underwent significant thrust forces and rotation at elevated temperatures. Figure 10-2 depicts different failure limit states of a typical steel shear connection at elevated temperatures that was simulated using finite element analysis (FEA).

In general, it has been shown experimentally that ductility is one of the most important aspects of a structural connection's resiliency to fire exposure. Moment connections designed with sufficient ductility (e.g., for seismic

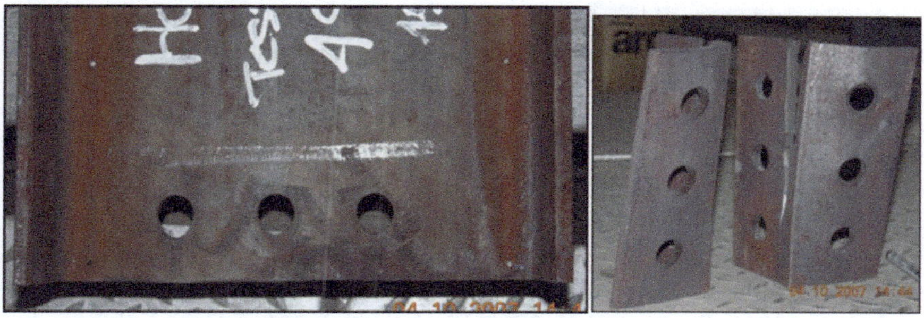

Figure 10-1. Web cleat connection failure at elevated temperatures.
Source: Photo by Ian Burgess.

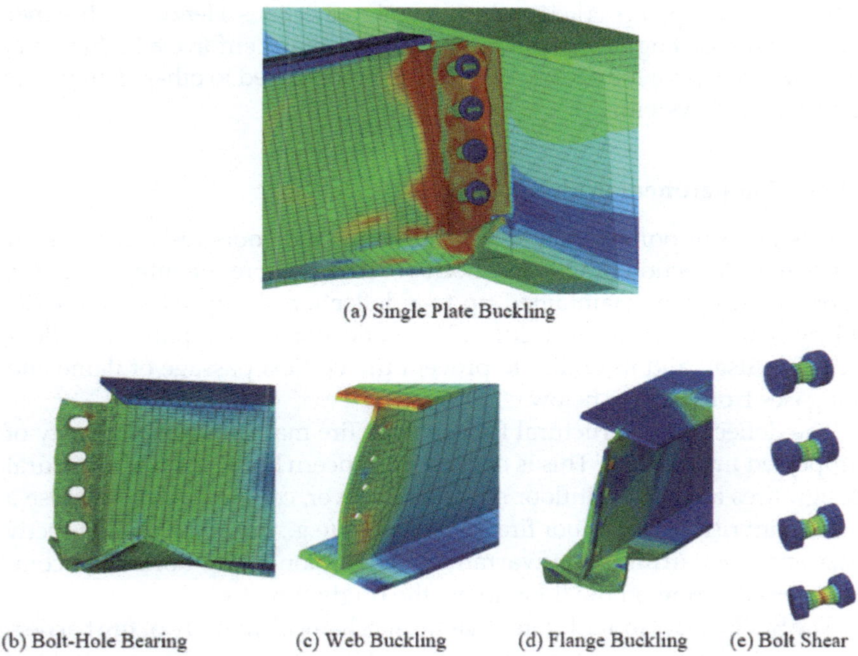

Figure 10-2. Different steel shear connection failure limit states (a–e) at elevated temperatures simulated using FEA.
Source: Image by Serdar Selamet.

design) are less likely to require additional moment capacity when contemplated for SFED. Generally, simple shear connections are not as robust and are more susceptible to local failure modes under fire exposure. For example, local buckling of a steel beam's lower flange may occur once the gap tolerance of the connection is exceeded because of large deflection of the beam under fire exposure. Also, local buckling of a beam's lower flange and web may occur once the shear connection develops large compression due to thermal thrust restraint and negative moment caused by the rotational restraint of the connection. Section 10.2.1 provides guidance for evaluating such local limit states. Ductile detailing provisions may be considered to establish whether these limit states will cause failure of the connection.

It has been demonstrated that structural connections with sufficient ductility are able to deform significantly and expansion forces are able to dissipate without collapse (Yu et al. 2009; Selamet and Garlock 2014; Selamet and Caner 2016; Burgess et al. 2012; Fischer et al. 2016). Generally, single-angle and double-angle connections have been shown to be the most robust traditional connection types for fire resiliency (Selamet and Garlock 2014, Selamet and Caner 2016). However, further research is required to establish generalized acceptance criteria for structural connections. Hence, the designer should exercise engineering judgment and perhaps enforce a higher level of conservatism for structural connections as compared to other elements of the structural system.

10.2.4 Compartmentalization

It is often important that structural floors and roofs resist fire spread vertically. Deflection limits have been observed where the integrity of the floor structure was maintained up to a deflection of approximately L/20 (Bisby et al. 2013, Wang et al. 2013). This deflection limit applies to the floor structure itself and its ability to prevent the vertical passage of flame and hot gases from the fire below.

The deflection of structural floors under fire may affect the integrity of supported fire barriers. This is often not a concern because most structural design fires are not multifloor events. However, certain cases may pose a significant risk of multifloor fire involvement (e.g., fire originating directly adjacent to an atrium) and warrant an evaluation of the horizontal compartmentalization on the floor above the originating fire.

The SFPE S.02 standard states that the mechanical integrity of fire barriers cannot be reliably predicted analytically (SFPE 2015). Hence, empirical test data may be required for critical fire barriers that are supported by the steel members to derive an appropriate deflection limit. The designer may also consult with manufacturers of fire resistance–rated wall products for more information. These effects are not explicitly contemplated by SFRD because the boundaries of the test furnace remain static during standard fire testing.

10.2.5 Egress Path Levelness

As discussed in Section 3.1, upholding the intended functionality of occupant egress systems is a minimum performance objective per ASCE/

SEI 7-16 (ASCE/SEI 2016). Accordingly, the designer should evaluate the fire-induced deflection of structural floors that support occupant egress routes. For such evaluations, the designer may apply a deflection limit of L/20, which is not expected to inhibit the ability of occupants to safely traverse the floor, despite visible deformations. Otherwise, a proposed deflection limit should be approved by the building authority and memorialized within a design brief document in accordance with the *SFPE Engineering Guide to Performance-Based Fire Protection* (SFPE 2007). The slope of ramps and other similar accessible egress features may increase under fire exposure. Accordingly, the designer should evaluate slopes (longitudinal and transverse) of accessible egress features under fire exposure and confirm that they do not exceed requirements of the governing accessibility code. Otherwise, accessible routes within public areas may not be traversable for handicapped occupants during a fire event.

10.2.6 Exit Stairways

As discussed in Section 3.1, upholding the intended functionality of occupant egress systems is a minimum performance objective per ASCE/SEI 7-16 (ASCE/SEI 2016). Accordingly, the designer should evaluate the stability of structural systems that support exit stairways under fire exposure. In addition, the designer should evaluate the integrity of fire barriers that comprise exit stairway enclosures. For ambient conditions, a deflection limit of L/480 is commonly applied to floors that support nonstructural components likely to be damaged by significant deflections (e.g., block partition walls) (CAN/CSA 2004). Because exit stairways are critical for occupant life safety in buildings, the designer may use this conservative deflection limit for evaluating the structural support of exit stairway enclosures under fire exposure. Otherwise, a proposed deflection limit should be approved by the building authority and memorialized within a design brief document in accordance with the *SFPE Engineering Guide to Performance-Based Fire Protection* (SFPE 2007) and perhaps supported by specific studies or test data.

10.3 COMPOSITE STEEL AND CONCRETE

10.3.1 Local Behavior

Under fire exposure, the shear connectivity between steel beams and the concrete slab may be partially or fully lost owing to localized concrete cracking or crushing and/or shear failure of steel shear studs. Currently, there is insufficient research data to derive generalized acceptance criteria for local concrete cracking/crushing in the vicinity of shear studs. Hence, the designer should exercise engineering judgment in cases where this behavior is critical or perhaps reference specific test results.

Any loss of shear connectivity between steel beams and the concrete slab under fire exposure could change the unbraced length of the steel beams. If this effect is deemed significant, the designer should either use conservative assumptions or evaluate the condition explicitly using FEA that is perhaps calibrated to relevant test results.

For composite floor systems with steel beams, the designer may apply the local behavior acceptance criteria from Section 10.2.1 to evaluate the steel members for SFED. In addition, the designer should properly assess the vertical shear reactions at the supports, including the available capacity of beam-to-girder connections and the concrete slab itself at elevated temperatures.

Acceptance criteria for shear stud failure are dependent on the complexity of the software package used and the extent to which shear stud behavior is captured. When FEA is used, Section 9.3.2 provides information on the temperature-dependent load-slip behavior of composite shear connectivity. Accounting for the uncertainty pertaining to localized concrete cracking/crushing, the designer may use this information to evaluate composite shear connectivity under fire exposure.

10.3.2 Gross Member

10.3.2.1 Simplified Analytical Tools. For composite systems, the designer may calculate temperature-dependent member resistance using applicable design methods in AISC 360 Appendix 4 (AISC 2016) or other approved methods. In either case, the designer should consider the reduction of the effective yield stress at elevated temperatures. In addition, AISC 360 (2016) permits the design for flexure in composite beams using either temperature-dependent reduced yield strength based on prescribed temperature variations over the cross section or temperature-dependent retention factors defined in Table A-4.2.4 based on beam bottom flange temperature. Overall, the designer should exercise engineering judgment to ensure that the simplified analytical tools used remain valid and accurately account for thermally induced effects for the temperatures expected.

When using simplified analytical tools, a demand-to-capacity ratio exceeding 1.0 does not necessarily indicate that a structural system would become unstable during a structural design fire, and stability may be demonstrated with a more advanced analysis. Further research is required to understand the anticipated performance of *over-capacity* composite steel floor systems when considering demand-to-capacity ratios calculated with simplified analytical tools under fire conditions. In these cases, more advanced methods of analysis (as discussed in Section 10.3.2.2) may be used to demonstrate structural stability during fire exposure.

In addition, the designer should demonstrate that all unbraced lengths are valid for the anticipated elevated temperatures to accurately account for global buckling and lateral torsional buckling that may occur with reduced member stiffness or strength. A bracing member is deemed to be effective if it can resist 0.02 times the factored compressive load at a brace point at ambient temperature. If this capacity is exceeded at elevated temperatures, the designer should consider such a brace point to be ineffective.

10.3.2.2 Finite Element Analysis. The acceptance criteria for FEA described in Section 10.2.2.2 are applicable to composite steel floor systems. At elevated temperatures, composite steel floors may develop tensile membrane action at high deflections, which provides an alternate load path for the supported load. Tensile membrane action may be dependent on the boundary conditions provided by surrounding structural framing. Also, predicting the ductility of steel reinforcement within a composite slab may be vital to assessing the development of tension membrane action because the reinforcement would carry tensile forces if the steel beams completely lose their load-carrying capacity at elevated temperatures.

To demonstrate the stability of this post-damaged state, the designer should use FEA. Thin composite slabs (e.g., 64-mm-thick concrete on 38-mm-thick deck) designed and constructed as per North American design guidance with unprotected secondary beams may not inherently be able to develop tensile membrane action without additional design considerations or full-scale testing (Wellman et al. 2011). Thicker assemblies such as those used in Cardington (i.e., 75-mm-thick concrete on 55-mm-thick deck) have been shown to develop tensile membrane action more favorably (Bailey et al. 1999).

For composite concrete floors, possible failure mechanisms related to tensile membrane action include crushing of concrete in the peripheral compression ring, rupture of steel reinforcing within the concrete slab, tensile cracking of the concrete slab itself, loss of vertical support at the perimeter of the slab, and connection failure to the intermediate steel beams that are supported by the deflected slab. Some of these failure mechanisms are illustrated in Figure 10-3.

Currently, there is no industry consensus for the interpretation of FEA output as it pertains to the assessment of possible failure mechanisms that pertain to tensile membrane action of a composite steel floor system. Illustrating this point, some case studies neglect to consider strains in the slab and focus on gross deflections (Wang et al. 2013), whereas others visually assess the distribution of strains for consistency along support lines but do not present quantitative acceptance limits for these strains (Flint et al. 2013). One such case study that quantifies strains due to tensile membrane action suggests that a rebar strain of 4% is deemed too high, necessitating an

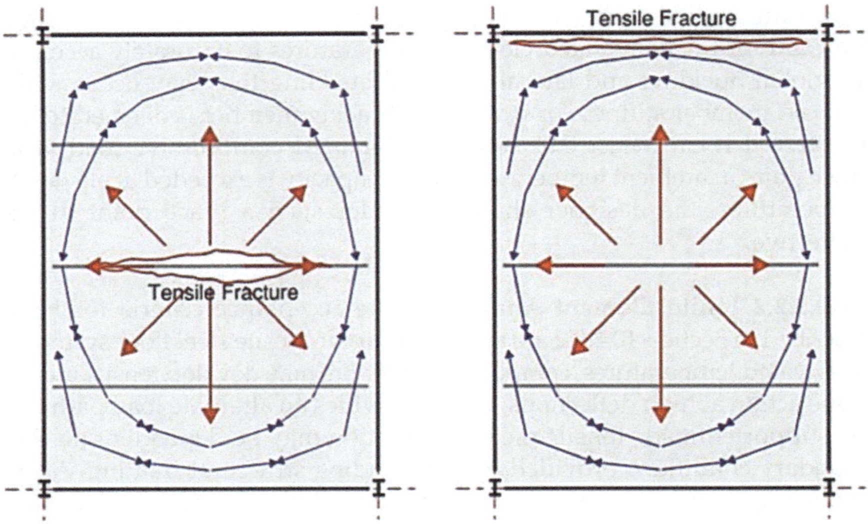

Figure 10-3. Specific failure mechanisms of tensile membrane action. Source: Image by Entuitive.

increase in reinforcing density (to control strains at 2%), although a reference for why this limit was deemed acceptable was not provided (Lamont et al. 2006).

It has been shown in the literature that additional reinforcement placed over supports should be explicitly captured in an FEA in addition to any evenly distributed reinforcing because it could limit strains at the perimeter of a slab panel and prevent runaway failure (Chang et al. 2014). Additional reinforcing over support girders is common in North American practice for unrelated reasons (e.g., control of concrete cracking). One acceptance criterion commonly applied across the published case studies is quantification of the ultimate slab deflections, which are related to overall stability as supported by the observations of the Cardington fire tests. As discussed in Section 10.2.2.2, the designer may consider this limit as L/20 for overall stability of a floor system (Wang et al. 2013). However, the designer should exercise engineering judgment to ensure that a FEA remains valid and accurately accounts for other thermally induced failure modes for the temperatures expected.

For an SFED that relies upon tensile membrane action of a composite steel floor system as a load support mechanism, it is especially important to have an independent third-party review conducted. This necessity is reinforced by the lack of industry consensus displayed in the most recent case studies previously mentioned. Also, failure mechanisms that can be accurately quantified are dependent on the software used, as some programs will make certain simplifications that must be properly reviewed for the given application.

A common area of simplification for FEA of composite steel floors is at the connection between the concrete slab and the steel beam, as some software programs treat this as a fixed connection, whereas others allow for a force- and temperature-dependent link to be modeled that captures slippage. The designer should represent force- and temperature-dependent behavior when using an FEA approach, unless it can be specifically demonstrated that a rigid representation is conservative for the application. In either case, the shear connector representation using FEA should be verified against available full-scale test data.

10.3.3 Connection Design

The acceptance criteria for FEA described in Section 10.2.3 are applicable to composite steel floor systems. The designer should demonstrate that beam and shear connectors are not compromised and that the slab reinforcement continuity to the cooler compartments remains intact. In all cases, the designer should evaluate the performance of steel connections as a critical aspect of maintaining continuity and stability in composite floor systems.

10.4 REINFORCED AND PRESTRESSED CONCRETE

Experimental studies conducted in the last few decades (Bailey 2002, Bisby et al. 2013, Yang et al. 2015, Gales et al. 2016) have highlighted the need for more research on reinforced and prestressed concrete structural systems that is relevant to SFED. The commentary herein recommends specific research areas that could help establish definitive acceptance criteria for realistically restrained reinforced concrete structures at elevated temperatures.

10.4.1 Concrete Cover Spalling

Historically, designers have relied on the inherent insulation characteristics of concrete cover to reduce the heating of tensile reinforcement under fire exposure. As discussed in Section 8.3.1.6, the designer should carefully consider the possibility of fire-induced spalling of concrete cover when conducting SFED (e.g., as shown in Figure 10-4). However, spalling is an apparently stochastic (erratic) and sudden phenomenon characterized by its multiparametric dependency upon various influencing risk parameters. Also, there is limited research on acceptable limits (e.g., temperature) for spalling (Maluk et al. 2016). Hence, further research studies are needed to develop appropriate acceptance criteria for concrete cover spalling. It is envisioned that such acceptance criteria would focus on quantifying the beneficial

Figure 10-4. Spalling of concrete cover from post-tensioned concrete. Source: Photo by John Gales.

effects of specific mitigation techniques, most notably the incorporation of polypropylene fibers into the concrete mix design.

Current Eurocode 2 (CEN 2004) recommendations specify limitations where spalling should be assumed based upon the moisture content of concrete by weight. Existing recommendations are 3%. However, it has been shown that spalling can still occur below this moisture content. Nonetheless, it would be conservative to assume a local loss of concrete cover to reinforcing bars and then to verify the reduced load-bearing capacity of the section. New spalling mitigation recommendations are being advocated for future versions of the Eurocode. Organizations such as the International Union of Laboratories and Experts in Construction Materials, Systems and Structures (RILEM) are initiating these discussions. For now, the designer should make conservative assumptions regarding the possibility of spalling and the beneficial effects of additives when conducting SFED.

10.4.2 Local and Global Behavior

As summarized by Bisby et al. (2013) and Gales et al. (2016), there are limited research and test data on the structural system performance of reinforced and prestressed concrete structures that are relevant to SFED. Accordingly, further research studies are needed to develop appropriate local and global behavior acceptance criteria.

Of the available studies that are relevant to SFED, it is observed that any out-of-plumb concrete column behavior and/or slab deflection should be carefully considered by the designer (Sharry et al. 1974). However, more research is needed to define acceptable limits, particularly when related to realistic restraining mechanisms in actual buildings. Hence, the designer should exercise engineering judgment and strive for conservatism owing to the lack of information pertaining to the local and global response of reinforced and prestressed concrete structures under fire exposure.

10.4.3 Analytical Tools and Analysis

If the designer can reasonably demonstrate that spalling would not present an issue for the structural design fires under consideration, methods described in Section 9.3 may be used to evaluate the structural performance of reinforced and prestressed concrete structures under fire exposure. However, the designer should use conservative judgments for aspects of the SFED that have significant knowledge gaps, as identified by Yang et al. (2015).

There are literature sources that provide acceptance criteria for reinforced and prestressed concrete structures under fire exposure (e.g., Concrete Society 2008). However, the designer should be aware that such sources do not consider realistic fire scenarios and/or they have limited applicability to SFED as discussed by Wang et al. (2013). Hence, the designer should exercise engineering judgment and strive for conservatism when applying analytical tools and analysis for reinforced and prestressed concrete structures under fire exposure. Encouragingly, international research is currently being conducted to address some of the knowledge gaps that exist.

10.5 TIMBER

Recently, project stakeholders in certain jurisdictions have advocated for the use of solid-sawn and engineered structural timber products for mid- to high-rise construction, which would not otherwise be allowed per SFRD. Moreover, there have been many recent research studies on the performance of timber structures under fire exposure. However, there is a very limited number of completed studies applicable to SFED. Hence, the development of definitive acceptance criteria for timber structures under fire exposure requires more research. This is particularly true for unprotected or exposed timber construction. Nonetheless, relevant sources and detailed commentary are provided herein in lieu of definitive acceptance criteria.

10.5.1 Gross Member

For heavy timber members, the amount of charring is a critical factor to consider when performing SFED. However, the designer may not be able to reliably predict the char rate for a given structural design fire (as described in Section 8.5.4). If the anticipated degree of charring can be reliably predicted, the designer should evaluate the load-carrying capacity of the unaffected section using conventional design equations. Also, the designer should assume that the char layer of a timber section does not have any capacity to support loading and can therefore be neglected for structural analysis purposes. Lastly, the designer should demonstrate that the unaffected section relied upon for loading-carrying capacity would not experience any adhesive breakdown, which may occur below the char layer, as discussed in Section 10.5.1.2.

10.5.1.1 Char Depth. Information on char rates of timber derived from standard fire testing (per SFRD) far exceeds that for realistic fire exposures. The designer should demonstrate that the selected char rate is conservative per the structural design fires under consideration. However, understanding the influence of varying fire conditions on timber charring rates is an ongoing research area. For instance, Hopkin et al. (2016) and Richter and Rein (2015) have examined the use of thermophysical and kinetic modeling to derive appropriate char rates for SFED.

In addition to the fully developed phase, the designer should consider the charring of timber structures during the decay phase of a structural design fire. However, there is little to no research on timber char rates during this phase of a fire. Hence, the designer should exercise engineering judgment and perhaps assume an unchanged char rate from the fully developed phase to remain conservative.

10.5.1.2 Adhesion Capability. Engineered timber products are made utilizing a number of engineered adhesives—mainly phenolic (thermosetting) polymers. These enable the layering of thin strips or smaller pieces of wood to construct engineered timber products such as cross-laminated timber (CLT), glued-laminated timber (glulam), and laminated veneer lumber (LVL). As timber chars under fire exposure, these layers of adhesive may also degrade to a certain depth below the char layer (e.g., as shown in Figure 10-5). Consequently, charred portions of timber elements may subsequently fall off (Frangi et al. 2009, Bartlett et al. 2015).

The designer should consider the thermal profile of timber elements under fire exposure and determine the degree and extent to which adhesive degradation may occur below the char layer. However, this is an ongoing research area, and caution should be exercised accordingly. Recent research has begun exploring the effect of adhesive deterioration beyond the char layer (Emberley et al. 2016, Quiquero and Gales 2017, Otto et al. 2017), which may lead to a further understanding of the load-carrying capabilities of the noncharred wood regions. This research is still too premature to specify an acceptable criterion as a limiting temperature or affected depth. Some researchers have proposed limiting the temperature of the

Figure 10-5. Laminated veneer lumber showing adhesive degradation. Source: Photo by John Gales.

adhesive layer below 200°C to prevent delamination of the timber (Osborne et al. 2012). Nonetheless, it is recommended that the designer prevent the heating of adhesive layers above ambient unless specific test data can be relied upon, to remain conservative.

10.5.2 Connection Design

There is limited research regarding the stability of modern timber connections under fire exposure. Accordingly, further research will be required to define acceptable deformations for modern timber connections under fire exposure. It is recommended that the designer prevent the heating of timber connections above ambient unless specific test data can be relied upon, to remain conservative.

REFERENCES

AISC (American Institute of Steel Construction). 2016. *Specification for structural steel buildings.* ANSI/AISC Standard 360. Chicago: AISC.

ASCE/SEI (Structural Engineering Institute). 2016. *Minimum design loads and associated criteria for buildings and other structures.* Reston, VA: ASCE.

Bailey, C., T. Lennon, and D. Moore. 1999. "The behaviour of full-scale steel-framed buildings subjected to compartment fires." *Struct. Eng.* 77(8), 15–21.

Bailey, C. 2002. "Holistic behaviour of concrete buildings in fire." *Proc., Institution of Civil Engineers—Structures and Buildings,* 152(3), 199–212.

Bartlett, A., R. Hadden, L. Bisby, and A. Law. 2015. "Analysis of cross-laminated timber charring rates upon exposure to non-standard heating conditions." *Proc., Conf. Fire and Materials,* San Francisco, February.

Bisby, L., J. Gales, and C. Maluk. 2013. "A contemporary review of large-scale non-standard structural fire testing." *Fire Sci. Rev.* 2(1), 1.

Bono, J. 1963. "Method for fire tests of floor and ceiling assemblies." *Symp., Fire Test Methods.* West Conshohocken, PA: ASTM International.

Burgess, I., J. Davison, G. Dong, and S. Huang. 2012. "The role of connections in the response of steel frames to fire." *Struct. Eng. Int.* 22(4), 449–461.

CAN/CSA (Canadian Standards Association). 2004. *A23.3: Design of concrete structures.* Toronto: CSA Group.

CEN (European Committee for Standardization). 2004. *Eurocode 2—Design of concrete structures. Part 1–2: General rules—Structural fire design.* EN 1992-1-2. Brussels: CEN.

Chang, I., G. Flint, and N. Butterworth. 2014. "Pros/cons of transverse rebar in structural fire response of a composite structure." *Proc., 8th Int. Conf. Structures in Fire,* Prague, Czech Republic: Czech Technical Univ.

Concrete Society. 2008. "Assessment, design and repair of fire-damaged concrete structures." *Technical Report 68.* Surrey, UK: Concrete Society.

Emberley, R., A. Nicolaidis, D. Fernando, and J. Torero. 2016. "Changing failure modes of cross-laminated timber." *Struct. Fire,* 643–649.

Fischer, E., K. Selden, and A. Varma. 2016. "Experimental evaluation of the fire performance of simple connections." *J. Struct. Eng.* 143(2), 04016181.

Flint, G., S. Lamont, B. Lane, H. Sarrazin, L. Lim, D. Rini, et al. 2013. "Recent lessons learned in structural fire engineering for composite steel structures." *Fire Technol.* 49(3), 767–792.

Frangi, A., M. Fontana, E. Hugi, and R. Jübstl. 2009. "Experimental analysis of cross-laminated timber panels in fire." *Fire Saf. J.* 44(8), 1078–1087.

Gales, J., K. Hartin, and L. Bisby. 2016. *Structural fire performance of contemporary post-tensioned concrete construction.* New York: Springer.

Gernay, T., A. Millard, and J. Franssen. 2013. "A multiaxial constitutive model for concrete in the fire situation: Theoretical formulation." *Int. J. Solids Struct.* 50(22–23), 3659–3673.

Gernay, T., and J. Franssen. 2015. "A plastic-damage model for concrete in fire: Applications in structural fire engineering." *Fire Saf. J.* 71, 268–278.

Hopkin, D., J. Schmid, and K. Friquin. 2016. "Timber structures subject to non-standard fire exposure —Advances and challenges." *Proc., WCTE World Conf. on Timber Engineering*. Vienna, Austria: Vienna Univ. of Technology.

Johnson, G., and W. Cook. 1985. "Fracture characteristics of three metals subjected to various strains, strain rates, temperatures and pressures." *Eng. Fract. Mech.* 21(1), 31–48.

Kanvinde, A. 2016. "Predicting fracture in civil engineering steel structures: State of the art." *J. Struct. Eng.* 143(3), 03116001.

Lamont, S., B. Lane, G. Flint, and A. Usmani. 2006. "Behaviour of structures in fire and real design—A case study." *J. Fire Protect. Eng.* 16(1), 5–35.

Maluk, C., L. Bisby, M. Krajcovic, and J. Torero. 2016. "A heat-transfer rate inducing system (H-TRIS) test method." *Fire Saf. J.* https://doi.org/10.1016/j.firesaf.2016.05.001.

Myers, A., G. Deierlein, and A. Kanvinde. 2009. "Testing and probabilistic simulation of ductile fracture initiation in structural steel." *John A. Blume Earthquake Engineering Center Technical Report 170*, Stanford Digital Repository. http://purl.stanford.edu/tn725xk1691.

Osborne, L., C. Dagenais, and N. Bénichou. 2012. "Preliminary CLT fire resistance testing report." *Point-Claire, Canada: Advanced Building Systems–Serviceability and Fire Group*, Point-Claire, Canada.

Otto, A., H. Todd, and J. Gales. 2017. "Mass heritage timber performance in fire." *Ann. Conf.*, Montréal, Canada: Canadian Society for Civil Engineering.

Quiquero, H., and J. Gales. 2017. "Comparing timber adhesive shear strength properties after fire damage." *Proc., 15th Int. Conf. on Fire and Materials.* Bromley, UK: Interscience Communications Ltd.

Richter, F., and G. Rein. 2015. "Numerical simulation of the charring of timber structures under nonuniform fires." *Proc., 1st European Workshop—Fire Safety of Green Buildings.* Brussels: European Cooperation in Science and Technology (COST).

Ryan, J., and A. Robertson. 1959. "Proposed criteria for defining load failure of beams, floors and roof constructions during fire tests." *J. Res. Natl. Stand., Sec. C*, 63(2), 1999.

Seif, M., T. McAllister, J. Main, and W. Luecke. 2016. "Finite element modeling of steel moment connections with fracture for structural fire analyses." *AISC Eng. J.* 53(1), 47–59.

Selamet, S., and M. Garlock. 2014. "Fire resistance of steel shear connections." *Fire Saf. J.* 68, 52–60.

Selamet, S., and B. Caner. 2016. "Fire resilience of shear connections in a composite floor: Numerical investigation." *Fire Saf. J.* 81, 97–108.

SFPE (Society of Fire Protection Engineers). 2007. *Engineering guide to performance-based fire protection*, 2nd Ed. Gaithersburg, MD: Society of Fire Protection Engineers.

SFPE. 2015. *Engineering standard on calculation methods to predict the thermal performance of structural and fire resistive assemblies*. SFPE S.02. Gaithersburg, MD: Society of Fire Protection Engineers.

Sharry, J., C. Culver, R. Crist, and J. Hillelson. 1974. "Military Personnel Records Center fire, Overland, Missouri (Part 1)." *Fire J.* 68(3)5–9.

UL (Underwriters Laboratories). 2011. *Standard for rapid rise fire tests of protection materials for structural steel.* UL1709. Northbrook, IL: Underwriters Laboratories.

Wang, Y., I. Burgess, F. Wald, and M. Gillie. 2013. *Performance-based fire engineering of structures*. Boca Raton, FL: CRC Press.

Wellman, E., A. Varma, R. Fike, and V. Kodur. 2011. "Experimental evaluation of thin composite floor assemblies under fire loading." *J. Struct. Eng.* 137(9), 1002–1016.

Yang J., M. Bundy, J. Gross, A. Hamins, F. Sadek, and A. Raghunathan. 2015. "International R&D roadmap for fire resistance of structures." *NIST/CIB Workshop (NIST SP 1188)*. Gaithersburg, MD: NIST.

Yu, H., I. Burgess, J. Davison, and R. Plank. 2009. "Experimental investigation of the behaviour of fin plate connections in fire." *J. Constr. Steel Res.* 65(3), 723–736.

INDEX

Figures are indicated by *f*; Tables are indicated by *t*